The Challenges of Agricultural Production and Food Security in Africa

The Challenges of Agricultural Production and Food Security in Africa

Edited by

Olusegun Obasanjo and Hans d'Orville

Routledge
Taylor & Francis Group
New York London

First published by Taylor & Francis Ltd

This edition published 2012 by Routledge

Routledge
Taylor & Francis Group
711 Third Avenue
New York, NY 10017

Routledge
Taylor & Francis Group
2 Park Square, Milton Park
Abingdon, Oxon OX14 4RN

THE CHALLENGES OF AGRICULTURAL PRODUCTION AND FOOD SECURITY IN AFRICA

This book was set in Times Roman by Hemisphere Publishing Corporation. The editors were Radhika Rao Gupta, Heather Jefferson, and Carolyn V. Ormes; the production supervisor was Peggy M. Rote; and the typesetter was Laurie Strickland. Cover design by Michelle Fleitz.

A CIP catalog record for this book is available from the British Library.

Library of Congress Cataloging in Publication Data

The challenges of agricultural production and food security in Africa / edited by Olusegun Obasanjo and Hans d'Orville.

p. cm.

Derived from the papers submitted, and the presentations made, at a conference convened by the Africa Leadership Forum in Ota, Ogun State, Nigeria in July 1989.

Includes bibliographical references and index.

1. Agricultural productivity—Africa—Congresses. 2. Food supply—Africa—Congresses. I. Obasanjo, Olusegun. II. Orville, Hans d', date. III. Africa Leadership Forum.

HD2117.C47 1992
338.1'96—dc20 91-46540
ISBN 0-8448-1724-4 CIP

Contents

Preface

The challenges of seriously confronting agricultural production and food security in Africa have become of strategic importance and urgency in the successful socioeconomic development and transformation of the African economies. Much remains to be done to create the incentives and structural context in which African agriculture can be truly modernized and its productivity enhanced. The ultimate objective must be to increase food productivity and food self-sufficiency with a view to turning Africa into a food exporter.

The agricultural and rural crisis besetting Africa today is the result of both policy failures and structural rigidities that inhibit access to and control of vital resources. The challenge of leadership in the agricultural sector in Africa is how to design and implement policies that will help induce growth and development in the sector.

It is against this background that the Africa Leadership Forum, in July 1989, convened a major conference to explore a range of parameters to be addressed in the formulation of successful policies. This volume is derived from the papers submitted and the presentations made at the conference, which was held in Ota, Ogun State, Nigeria.

In the Appendix to this book, the reader will find the main conclusions and recommendations that emerged from the conference, which have since been widely distributed among decision-makers in Africa and among concerned international organizations. Proceeding from an analysis of what went wrong in the past, the conclusions highlight a series of actions that must be taken in a number of areas, including women in agriculture, small-scale farmers, agribusiness, subsidies, human capital, linkages between international and national research, and energy. The substantive contributions from which these conclusions were derived served as the basis for this text.

In the Introduction, Olusegun Obasanjo, Chairman of the Africa Leadership

Forum, reviews in a *tour d'horizon* the manifold problems and shortcomings of African agriculture and suggests a number of measures that might be taken to alleviate present problems.

The well-known Indian scientist M. S. Swaminathan examines in detail the present situation and discusses the nature of challenges, including social factors, economic factors, energy, employment, ecology, and social engineering. He subsequently elaborates several strategies corresponding to the various ecological zones of Africa. One key aspect of this chapter focuses on water as a key determinant of the agricultural future. Likewise, he discusses the interrelationship between ecological and food security and biotechnology and argues that strategies should be oriented to specific target groups, e.g., sustainable development systems for small farmers or women. Swaminathan concludes with an outline of new frontiers and an agenda for action.

This overview chapter is supplemented by a detailed study of Africa's food security situation by the Secretariat of the United Nations' Economic Commission for Africa (ECA). The ECA study develops the concept of food security and offers an assessment of the current food security situation in all African countries. From there, it develops the resource management challenges for sustainable food production in Africa and finally raises the issue of food accessibility.

A contribution by Idris Nur, of the Secretariat of the Organization for African Unity, deals with dietary patterns in Africa. Knowledge of these patterns is of crucial importance in determining the provision of adequate food to the population in the various countries and ecological zones.

After these more general chapters, another series of chapters discusses specific aspects of the agricultural practices and problems in Africa. Thomas R. Odhiambo of Kenya, Director of the International Centre of Insect Physiology and Ecology (ICIPE) and Chairman of the African Academy of Science, analyzes from a scientific viewpoint the problems and prospects of the practice of agriculture in Africa. This chapter discusses in detail pest-management innovations and especially the adoption of an integrated pest-management system in tropical Africa as a critical means to help ensure, year after year, the harvest of a reasonable yield of a farmer's crop.

This is followed by chapters dealing specific aspects of the practice of agriculture in Africa. Mark Chona of Zambia and the National Farmers Association of Zimbabwe expound on the farmer's viewpoint and his or her problems. Mary Okelo of Kenya discusses the role and problems of women in the agricultural context.

George Moody-Stuart, an executive with Booker plc. of the United Kingdom, shares his vast experience in agribusiness, i.e., the agriculture organized on a large scale. Proceeding from the question of whether Africa needs agribusiness, he specifies what agribusiness needs from government and what governments, in turn, can get from agribusiness. Pierre-Victor Mpoyo, an African entrepreneur, then outlines his perception of the needs of African agriculture.

After this detailed focus, the next chapters deal with international organizations active in the agricultural field and their respective policies and approaches. Ketema Yifru of Ethiopia describes the perspective of the World Food Program (WFP); Leopold Wantisse Siry of Burkina Faso discusses the overall approach of the Food and Agricultural Organization of the United Nations (FAO) in formulating national food security programs; and Eugene R. Terry of Sierra Leone outlines the role and strategy of the West Africa Rice Development Association (WARDA), a regional organization engaged in the formulation of a strategy for rice research.

To ensure ecologically sustainable agricultural development in the future, environmental and ecological considerations will be crucial. Thus, the contribution of Edward S. Ayensu of Ghana deals with balancing conservation and development; Paul Harrison, the British author of the bestseller *The Greening of Africa,* tells What is Going Right in African Agriculture, including an emphasis on successful soil conservation programs.

The successful experience of the Republic of Korea in developing its economy into one of the most successful in Southeast Asia has attracted the interest of many observers. Can its experience and development path be emulated, and are there thus lessons that may be learned by African countries? Jai-Suk Chung, a former cabinet minister of the Republic of Korea and now a professor at Hankuk University, presents the Korean Experience: Development, Food, and Agricultural Sufficiency.

The volume concludes with former U.S. President Jimmy Carter describing his international activities in the agricultural field, mainly aimed at inducing a green revolution for Africa.

Olusegun Obasanjo
Hans d'Orville

Foreword

by Jimmy Carter

For perhaps the first time in history, we have at hand the resources and the opportunity to achieve food security throughout the world. Hybrid seeds and varieties, appropriate agricultural technology, medicines to eradicate debilitating parasites, and effective yet inexpensive fertilizers: all the essential tools have been developed and only need be put in the hands of farmers.

Through the Carter Center's Global 2000 program, we have been trying to do just that; but along the way we have learned lessons that humbled without daunting us. We have come to understand that food security is much more than adequate agricultural yields, that it brings into play such interrelated problems as health, nutrition, national infrastructure, the environment, civil peace, and political leadership. It is not enough just to help farmers produce record crops of food grains. They need a support system that ensures the grain is not lost to spoilage or pests, allows the grain to be transported to market, guarantees a decent price for the farmer at market, and provides for the grain to be processed into flour affordable to the poorest family.

The world's donor agencies and nations see the great opportunity at hand. Africa's leaders are in a position to make the policy decisions required to deal with the complex issues behind attaining food security. For that reason, the Africa Leadership Forum genuinely excites me. Here is a convergence of concerned leaders who are determined to make their peers aware of the problems facing the African continent. We find great hope in the future—because an organization like the Forum is confronting the last major obstacle to attaining food security: leadership.

The Africa Leadership Forum is one of the critical new developments that offers hope for a solution to Africa's problems.

Chapter 1
Introduction

Olusegun Obasanjo

Time was when the word "crisis" might have evoked the response that its correct usage demanded. I believe that the origins of the word lie in pathology and in its original pathological sense it meant that stage in the development of a disease where the patient either recovered or succumbed. Inevitably, it was a stage that concentrated minds wonderfully—the mind of the patient himself in the first place, but also the mind of the physician and those of the patient's friends and loved ones. Today, the word "crisis" seems to have lost practically all its evocative connotations and has become just another cliché, like any other. After the years of protected drought and famine in Africa, after Bhopal, Chernobyl, the Rhine chemical spillage, or, to come nearer home, the mud slide in Colombia or Lake Nyos, the world appears to have become inured to the realities of crises. Not that long ago all these disasters would probably have been perceived as nothing short of the opening of the gates of hell. Today, I suspect that a growing number of people are coming to see and perhaps even accept them as part of the changing order of things.

Yet I can think of no other term that fully covers all the facts of the present state of African agriculture. However, I use the term in a slightly different sense. I am informed that the Chinese equivalent of the term "crisis" is designated by two words, one that means "danger" and the other "opportunity." It is in this dual sense that I propose to treat the crisis of African agriculture and the associated issue of food security. But before dealing with the present dangers and opportunities of African agriculture, I would like to cast a retrospective glance at sub-Saharan Africa before the arrival of European technology. Edmund Burke once said that whenever the Spanish Council of Ministers of a certain epoch were at a loss on some point of policy or other, they would say, "Let us consult the wisdom of Philip II." In like manner, as we in our turn cast about for lasting solutions to our agricultural and other problems, we would do well to begin by consulting the wisdom of our ancestors.

Precolonial Africa was preeminently an agricultural world. This was true of the Sudanic Empires of medieval Islam as it was of the nineteenth-century states of the Yoruba, the Asante, and the Hausa-Fulani people. It is true that trade played a very significant role in the fortunes of some of these polities. But trade, commerce, and industry were usually secondary to agriculture. More often than not, the long-distance trader or the blacksmith was himself no more than an off-season farmer who turned his hand to trade or the forge in the fallow months of the dry season. Agriculture was king and the agricultural cycle dictated the rhythm of social life. Nor was it merely subsistence agriculture in all cases. In the grasslands of the Savanna, where cereal culture was dominant, agriculture both met domestic needs and generated a surplus for the trans-Saharan export trade.

Let me say at once that traditional African agriculture knew neither the wheel nor the cart although knowledge of the use of iron went back several centuries before Christ. But what it lacked in technological implements it made up for in other ways. Over the centuries, through trial and error, the precolonial African farmer became the repository of a deep and profound ecological knowledge that in some respects amounted to near-mastery of his environment. He also knew the names of practically all the plants that grew around him along with their various uses. He knew what soils were good for what crops, the best time of the year to plant them, and how to nurture them. But there was something else that distinguished traditional agricultural practices in Africa and that has an immediate relevance for us today. Those farmers did exploit the land for a livelihood but its use of this resource was governed by clear and unbreakable rules. In the first place, customary law upheld a sharp distinction between ownership of the fruits of the land on the one hand and ownership of the land itself on the other. Individual ownership of land with an unfettered right to dispose of it as the individual sees fit was inconceivable in the traditional scheme of things. Indeed, not even the living members of the lineage in their corporate personality could validly claim to own the land in the African understanding of the concept of ownership. The living were only custodians of a patrimony that, in the formulation of the Ghanaian chief, Sir Nana Ofori Atta I, belonged to the dead, the living, and the generations yet unborn. This, I believe, was traditional society's insurance and assurance that the common resources of the earth would remain truly common and not be exploited. That responsible use of the earth's resources we may today call "sustainable development." "Sustainable development is progress that meets the needs of the present without compromising the ability of future generations to meet their own needs" (World Commission on Environment and Development, 1987).

There was another strength of traditional African agriculture that we should recall, not in any spirit of self-indulgent romanticism but as a source of inspiration. For all its limitations, precolonial African agriculture was responsive to new opportunities and stimuli from without. Consider for example the case of palm oil. The end of the slave trade at the beginning of the nineteenth century touched off a huge demand for palm oil, especially in Britain. Without any of the extension services and other inputs that were later to be introduced by the European presence, Nigerian farmers, relying on the resources of traditional agricultural methods, tried to meet the demand. In 1830, they sold 10,000 tons of oil to Britain. By 1855, the figure had risen to 40,000 tons, and increased steadily thereafter. Or take the case of cocoa in Ghana. Its introduction into the country in the second half of the nineteenth century was largely the result of local initiative exploiting a perceived opportunity. Thanks to the dynamism of its farmers, Ghana had by the end of World War I established itself as the world's premier cocoa producer.

What, we must now ask, were the underlying factors accounting for the viability and adaptability of traditional agriculture in sub-Saharan Africa? Clearly they were many, but I believe that the two factors that provided the foundation for everything else in the success of traditional methods of agriculture were the communal ethic and profitability.

In agriculture as in much else in traditional African society, no man was an island entire to himself. Everyone was a part of the main, and the communal ethic, so emblematic a feature of traditional society, was nowhere better exemplified than in agriculture. At every critical stage of the agricultural process, whether it was clearing the forest to make a new farm, helping to weed a neighbor's farm, or bringing in the harvest, the communal spirit was always in play. Almost certainly, communalism as a mode of social organization was initially enjoined on traditional society by force of circumstances, but once in place it was consolidated by its own invaluable merits until it became the defining characteristic of that society. If then in our agricultural endeavors we have stood on the shoulders of many generations and still failed to "see" much further, the explanation surely lies in the fact that we have not fastened sufficiently onto what was in effect the engine of the success of traditional agriculture.

I shall develop this point further in my conclusions, in which I hope to adumbrate a few ideas on the way forward. Let there be no misunderstanding as to what it is that I am driving at here. Obviously, no historical experience is entirely repeatable and, if only for this reason, I hope I will not be construed as advocating a return to the pastoral societies of our forefathers. Neither am I a Romantic hankering after a pantisocracy. I hope to show that the moral of the communalism of our ancestors points to the future, not to the past.

THE DANGER

The chapter by Dr. Swaminathan provides a very detailed exposé of the major issues of African agriculture and some indication of the way forward; no useful purpose will be served by duplicating that effort here in however summary a form. However, I will point up some of the salient features of the crisis of African agriculture as a backdrop to a consideration of what might be done to contribute to a resolution of that crisis.

All the reviewed sources are agreed that over the past 25 years or so, African agriculture has either suffered a net retardation or a consolidation stuntification. Statistics may vary depending upon which international agency compiled them and when they were complied, but the margin of difference is usually such as to leave little doubt that agriculture, like other sectors of African society, is trembling on the brink and that unless something happens and happens sooner rather than later, most of sub-Saharan Africa at least will go over that brink. Because no one can seriously question the integrity or credibility of the report of the independent World Commission on Environment and Development (also called Brundtland Commission after its Norwegian chairman), let me reproduce its schematic perspective on agricultural development in Africa as at 1987:

> "a drop in per capita food output of about 1 percent a year since the beginning of the 1970s;
> a focus on cash crops and a growing dependence on imported food, fostered by pricing policies and foreign exchange compulsions (and manipulations);
> major gaps in infrastructure for research, extension, input supply, and marketing;
> degradation of the agricultural reserve base due to desertification, droughts, and other processes;
> large untapped potential of arable land, irrigation, and fertilizer use."

The Commission's findings on comparable regions of the world were very revealing. Take for example its perspective on South and East Asia. There, it was a case of

> "increased production and productivity, with some countries registering grain surpluses;
> rapid growth in fertilizer use in some countries and extensive development of irrigation;
> government commitments to be self-reliant in food, leading national research centres to development of high-yielding seeds, and the fostering of location-specific technologies;
> little unused land, and extensive, unabated deforestation;
> growing numbers of rural landless."

Even so the reality is still a great deal graver than the Commission's shorthand generalizations convey.

The gravity is particularly evident in the food situation. Since the end of the 1960s, per capita food production in Africa has declined by no less than 20 percent and it is estimated that if this trend continues, by the year 2000 most of sub-Saharan Africa will face severe food shortages. Already in 1983–1984 the region spent some US$15.2 billion on food imports and, based on present trends, it is projected that by the year 2010 food imports for the whole continent will account for US$28.5 billion against anticipated total agricultural export earnings of US$12 billion.

Against such a background, the concept of food security assumes an urgency all its own and even undergoes a slight variation in meaning. If the idea of food security is to have any meaning for the ordinary man and woman in Africa, if it is to be no more than a pair of words, then it would have to proceed in the manner of charity, that is to say, it would have to begin at home. It must mean first and foremost an adequate and balanced meal for the household, not just for a week, a month, or even a year, but for years, for a lifetime. It will then need to go beyond the household and encompass the entire country and then the region. How this is to be achieved is one of the most pressing challenges facing Africa's leadership today.

Declining food production is sometimes no more than the other side of the coin of a degraded environment. Until the late 1960s, the tropical rain forests of most of sub-Saharan Africa were still fairly intact; then at the turn of the 1970s accelerated deforestation began in earnest. By 1981 an FAO-UNEP report had established that Africa was losing 1.5 million hectares of forest a year—an area about half the size of Rwanda. In 1980 it was estimated that Liberia was logging 80,000 hectares of its remaining 900,000 hectares of primary forest. Cote d'Ivoire, the largest exporter of timber in Africa, saw its forest area decline by two thirds in 20 years, from 12 million hectares in 1956 to 4 million in 1977. In Ghana, the dense tropical forest is rapidly becoming a thing of the past. The information on the trends of the past nine years is patchy, but even without it we can be pretty sure that deforestation has continued apace.

In the same way as human predation converts the rain forest into grasslands, the savanna in its turn becomes desert. In the first half of the 1980s it was estimated that as much as 742 million hectares of land in Africa (equivalent to 26 percent of the continent's total land area) was undergoing desertification, with the Sahara alone expanding at the rate of 1.5 million hectares a year. This relentless greying of the continent is one of the factors at the root of the current socioeconomic crisis. Other associated causes of environmental degradation, such as overgrazing and overcultivation, have also contributed to the problems of agriculture and food production.

There is no doubt that man's exploitation of nature has contributed considerably to creating a world of disappearing forests and advancing desert. But that is by no means the whole story. In Africa other compulsions, more human and therefore in a sense, excusable, have also been at play. Many trees and thickets have been cut down in the Sahel not for export but either to meet the family's domestic fuel needs or to make a new farm. This means that with each increase in Africa's population whole swaths of forest and grassland are put at risk.

However, Africa is not overpopulated in terms of its land area but because of its resource and management capacity. With the current estimated growth rate of 3.2 percent a year, Africa has the fastest-growning population in the world. There is an imbalance between the carrying capacity of agriculture and the population growth rate; this imbalance is caused not so much by excess population but by the current low levels of agriculture technology in Africa. The World Bank (1985) has projected a population of 529 million for Nigeria by the year 2035. At the current rate of economic performance, this would be a disaster; averting such a disaster would require both technical and political efforts.

THE OPPORTUNITY

From the collapse of commodity prices on the one hand to the relative fall in the production of export commodities on the other, from the protracted drought and consequent famine of the 1970s and the early 1980s, from the debt overhang and the related decay of so much of our already exiguous socioeconomic infrastructure, one fact has now emerged: we in Africa are engaged in a race against catastrophe. Needless to say it will be protracted and painful and will stretch all our resources, human and material, to their limits. But we embark on that race confident in the knowledge that we have what it takes to win it.

In spite of our wholly censurable profligacy in the exploitation of our ecological resources, Africa still has enough land to feed its present and future generations for a long time to come. As Dr. Swaminathan (chapter 2, this volume) has usefully pointed out in his study, land is still relatively plentiful in Africa: only 25 percent of the potentially arable land is under cultivation in contrast to over 70 percent in Asia.

But land provides no more than the foundation; for its optimal exploitation we will need to call on the combined knowledge of our peasant farmers and research scientists. With the deepening of our crisis, it has now been grudgingly conceded that the traditional African farmer is not after all the superstitious ignoramus that officialdom, colonial and postcolonial, presumed him to be. Today the small peasant farmer's inherited experience and knowledge are seen as assets that still have to be mobilized in the service of development. But we shall not have the full benefit of that science unless our research scientists and policymakers learn to work in cooperation with the peasant farmer. Happily there is evidence that this is beginning to be the case; there is even an acceptance that local farmers often know more than the experts.

The aggravation of the crisis should also serve as an opportunity for the acceleration of much-needed and, in some cases, long overdue reforms in other areas. I believe that there is a place in contemporary Africa for something of the communal spirit of traditional African society. In the past I have warned that no

African country working alone stood a gambler's chance of resolving its economic and other problems within the ambit of its own frontiers and that redoubled efforts were necessary for regional and ultimately continental economic integration.

The agricultural crisis calls for the same rationality. Drought is not the problem of one country; it afflicts many from Mauritania to Ethiopia and from Chad to Mozambique. The encroaching Sahara affects not only Burkina Faso but the entire ECOWAS region and beyond. Under such generalized conditions of disaster, it would be quixotic for any one country to think that it can find salvation outside the fold. Indeed under such conditions, economic cooperation and integration, once a desideratum, becomes a pressingly urgent imperative. Communalism lives on but it now proceeds by the name of economic integration.

One practical proposal could be based on the precedent of the 1989 Antarctic Treaty: the possibility that all the African countries directly affected by the seemingly inexorable advance of the Sahara come together in some form of international authority to institutionalize the effort to stem the Sahara's advance. If practicable, such an organization would have incomparable advantages over the present isolated and woefully inadequate efforts of the affected countries to fend off the desert. Friends of Africa and friends of the earth could closely cooperate and collaborate with such a body.

We have all heard the insistent call for African governments to devise appropriate agricultural policies as a way out of the present swamps and shallows. Although appropriate policies must be formulated, even then we would still be no near the journey's end. The bane of African agricultural policy remains the howling emptiness between word and deed. In other words, more often than not it is a problem of implementation and this problem cannot be satisfactorily resolved except in the context of a general overhaul of the machinery of government.

REFERENCES

World Bank, "Desertification in the Sahelian and Sudamian Zones of West Africa" 1985.

World Commission on Environment and Development. *Our Common Future* (New York: Oxford University Press, 1987).

Part One
Overview

Chapter 2
Agricultural Production in Africa

M. S. Swaminathan

WHERE ARE WE NOW?

In the Arturo Tanco Memorial Lecture delivered in 1987, General Olusegun Obasanjo said, "the economic future of Africa depends on the future of agriculture." This point was also emphasized by President Robert G. Mugabe in his Africa Prize for Leadership address delivered in London on January 30, 1989, in the following words:

> Agriculture will continue to be the backbone of Zimbabwe's economic development for the foreseeable future. This section accounts for nearly half of our foreign exchange earnings. While in 1980 the exportation of some of our surplus accounted for Z\$282 million, this contribution had risen to nearly Z\$1 billion in 1988.

In other words, agricultural progress is the key not only to food security but to rural and national prosperity. This is why in my Aggrey-Fraser-Guigesberg Lecture delivered at the University of Ghana in 1981, I pleaded for a wider concept of food security. A national food security system can be built only on the twin foundations of ecological security and employment guarantee (Swaminathan, 1981). In the first Arturo Tanco Memorial Lecture delivered in 1986, I stressed the need for building sustainable national nutrition security systems in all developing countries. I defined "nutrition security" as physical and economic access to balanced diets and safe drinking water for all children, women, and men at all times. This is absolutely vital for ensuring that all the citizens of a country have an opportunity to fully express their innate genetic potential for physical and mental development. In the Andrew Sharman Lecture of the Royal Society of Arts delivered in 1985, I stressed that, to be sustainable, development must also be equitable. There is no point in talking about a better common future for humankind unless we are also committed to the cause of promoting a better common present.

It is now widely acknowledged that while current food security problems in Africa clearly originate in part from the stagnation or decline in food production, ensuring food security also implies the provision of what Amartya Sen calls "food entitlements" such as land, credit, income, and family support systems (Sen, 1981). The example of Botswana in introducing a multipronged strategy involving cash and food for work programs, supplementary feeding for underweight children, school feeding programs, and development of irrigation projects is often cited as an example of tackling concurrently the production and distribution aspects of the hunger problem.

A recent report titled *The Challenge of Hunger in Africa—A Call to Action* (The World Bank, 1988) lays stress on a multipronged strategy for dealing with the problem that more than 100 million people, i.e., a quarter of Africa's population, cannot pursue an active working life because they do not have enough to eat. The World Bank paper deals in detail with the quantitative and qualitative dimensions of the food insecurity problem in Africa and offers useful suggestions for a "mid-course correction" in current attitudes and approaches to food security problems in African nations.

A publication titled *Compact for African Development* (1985) and another paper titled *Transforming African Agriculture* both by Carl K. Eicher (1986) also contain useful insights and suggestions. The African food security challenge has led to the rapid growth of a vast library and an eloquent vocabulary generated by eminent scholars, social workers, and bilateral and multilateral donors, all of whom are concerned with the question, "Can Africa double the rate of growth of agricultural production over the next 15 to 20 years to keep up with population growth rates of 3 to 4 percent?"

One example of the new awareness of foreign experts on Africa is the realization that women's economic contributions to households appear to be even more important in periods of economic contraction than in periods of economic prosperity. Their contributions are also more critical in low-income than in high-income families and are central to ensuring the nutritional welfare of infants in poor families. Nearly all experts now agree that only a sustained process of agricultural diversification to provide higher incomes and more job opportunities in rural areas and the building of grassroots institutions to enable the rural poor to take advantage of the development opportunities will help to provide lasting solutions to the food insecurity problem. In this context, I recommend the book *Strengthening the Poor: What Have We Learned?*, edited by John P. Lewis (1988).

The report of the president of the World Food Council (WFC) titled *The Cyprus Initiative Against Hunger in the World* presented at Cairo in May 1989 describes the nature of the global food security challenge as follows:

> The number of hungry people increased 5 times faster in the 1980s than in the previous decades. There were more malnourished children in 1984 than in 1975. Although the world is now feeding one billion more people than at the time of the 1974 World Food Conference, the objective of assuring food security for all people has been moving further away.

As shown in Table 1, the number of undernourished in 1983–1985 was great not only in Africa but also in Asia and the Far East.

The president of the WFC concluded that the deterioration of the human condition in terms of hunger, malnutrition, and poverty will not diminish in the prevailing policy environment. On the contrary, with population growth, ecological deterioration, and a mixed outlook for the world economy and the economics of developing countries, the challenges of dealing with hunger and poverty will increase. Obviously, "a more of the same" approach is not going to reverse current trends and other avenues should be considered.

THE WAY AHEAD

Having concluded that if agriculture goes wrong nothing else will go right, how can we make agriculture go right? Example of countries such as China and India, which have been able to enhance agricultural production to the extent needed for national self-reliance in food output, indicate that farm women and men will be able to help the country to achieve its food security goals provided that they are enabled to do so. The Lagos Plan of Action adopted in April 1980 established as Africa's primary objective the alleviation of mass poverty and improvement in the people's standard of living. Subsequent discussions and meetings have dealt with why the goals of the Lagos Plan of Action were not realized during the 1980s.

The Economic Commission for Africa released on July 6, 1989, a proposal entitled *African Alternative Framework to Structural Adjustment Programmes*

Table 1. Numbers and percentages of undernourished in Africa, the Far East, Latin America, and the Near East

	Undernourished	
Region	Number (in millions)	Percentage of population
Africa	140	32
Far East	291	22
Latin America	55	14
Near East	26	11
Total	512	

Data given for 1983–1995.

for Socioeconomic Recovery and Transformation. The alternative blueprint for economic revival takes into account the following facts:

1. Per capita output declined from $752 in 1980 to $641 (in constant dollars) in 1987;
2. Per capita incomes declined in the same period by an annual average of 2.6 percent;
3. Africa may begin the next millennium with a greater proportion of its population being illiterate and unskilled than it did at the beginning of the postindependence era in the 1960s.

The African alternative proposal places a high priority on attaining food self-sufficiency. The various measures suggested for this purpose include:

1. Allocation of 20 to 25 percent of the total public investment to agriculture;
2. Land reform;
3. Enhancing the role of women;
4. Technological upgrading of the food production and processing sectors;
5. Minimum price for food crops;
6. Reorientation of credit policies to favor food production and the manufacture of essential goods;
7. Strengthening of rural financial institutions;
8. Greater focusing on African markets;
9. Building and managing strategic food reserves.

Thus, the African alternative places considerable stress on the development of mutually reinforcing packages of technology, services, and public policies. Above all, it emphasizes the need for local solutions to local problems.

Prescriptions for reversing current trends are many, resources are scarce, and time is running out. The biological impoverishment of the countryside and a continuous diminution in the biological potential of the land are compounding the difficulties arising from the already serious constraints confronting small farm families. Fortunately, land is not the most limiting factor in Africa because the continent occupies 11.70 million square miles of land as compared with 11.06 million square miles of the combined area of the United States, Western Europe, and China. Also in Africa only 25 percent of the potentially arable land area is now under cultivation as compared with over 70 percent in Asia. Africa is also fairly well endowed with good national and international research centers and training institutions. What then is the problem?

NATURE OF CHALLENGES

Social Factors

There are several social factors that contribute to the problem. For example, the average size of a farm holding is decreasing and there is a heavy overgrazing in some areas. The number of landless labor families is increasing and women usually perform jobs that involve much drudgery and little remuneration. Youth constitute the majority of the rural population and the educated among them are not satisfied either with the health care and educational opportunities available in villages or with the prevailing opportunities for skilled employment. Consequently, they tend to leave the villages to go to cities and towns.

Economic Factors

Small farms are affected more by the cost–risk–return structure of farming than are the more well-to-do sections of the farming community. Stability of income for small farmers is as important as yield per hectare. There is a mismatch between production and postharvest technologies, particularly with regard to perishable commodities and so knowledge-intensive techniques that can reduce the cost of production without reducing yield become particularly relevant. Also, income from farming alone is not adequate for a family to achieve a better quality of life: farm income will have to be supplemented by off-farm income.

Energy

Without adequate energy sources productivity and efficiency cannot be improved. The farm sector must have more energy if the efficiency of farm operations at the production and postproduction stages is to undergo marked improvement. Energy is particularly needed in dry and drought prone areas, both for the timely preparation of land for sowing and for taking advantage of groundwater resources. An integrated energy supply system involving both renewable and nonrenewable sources of energy needs to be developed for every agroecological region.

Employment

The farm sector plays a dominant role in providing employment. In order to attract and retain educated youth in agricultural occupations, farm technologies should be both intellectually satisfying and economically rewarding. A knowledge-intensive agricultural system that can also generate considerable employment in the secondary and tertiary sectors will be necessary to create more skilled jobs for rural men and women. An integrated approach to on-farm and off-farm employment and to blending traditional and frontier technologies is urgently needed. For this purpose, the ongoing Farming Systems Research (FSR) program will have to be modified and converted into a Rural Systems Research (RSR) program specifically designed to generate better growth linkages (Swaminathan, 1988).

Ecology

In the ultimate analysis, only ecological security can safeguard both food and water security and livelihood security. The conservation and sustainable management of land, water, flora, fauna, and the atmosphere are essential for safeguarding the future of African agriculture. The unique coastal, mountain, wetland, and dry-land ecosystems of Africa need urgent attention from the point of view of conservation and rehabilitation. Sustainable land use systems will require detailed attention with reference to the following three categories of land:

1. Land that needs to be conserved in its pristine purity because of its rich endowments of forests, wildlife, and other forms of biological diversity;
2. Land that is degraded and needs to be upgraded through appropriate techniques of restoration ecology. These lands are usually referred to as "wastelands";
3. Land that can be subjected to sustainable intensification through scientific agriculture and aquaculture.

Social Engineering

In view of the small size of holdings and the limited financial and technical resources available to small farmers and agricultural labor families, it is essential that the organizational aspects of small farm agriculture receive adequate attention. A small farm has great potential for intensive agriculture but a small farmer is usually confronted with serious economic problems arising from the cost and risks involved in intensive farming. Only by helping small farmers overcome these economic and technological problems can the potential of a small farm be realized. Also knowledge-intensive and ecologically sound agricultural practices, such as scientific water management, integrated pest management, and improved postharvest technology, require cooperation among a group of farmers living in a village or watershed or the command area of an irrigation project. The full potential of small farm agriculture can be realized if individual initiative, group cooperation, and government support can be promoted in a mutually reinforcing manner. Group insurance schemes, group subsidy, group-oriented extension, and other services and incentives for group cooperation need to be devised and introduced soon if we are to maximize the efficiency of small farm agriculture. Organizational structures that can promote decentralized production supported by a few key centralized services should be developed; these will also help to generate more skilled jobs in the service sector in rural areas. Special attention is needed in the case of perishable commodities for promoting producer-oriented marketing. For example, it would be useful to promote the organization of horticulture estates, poultry estates, and aquaculture estates to provide the needed infrastructure of facilities and services to small producers of vegetables, fruits, flower, eggs, poultry, and fish. Social engineering techniques should ensure that new technologies reach their intended targets and at the same time promote the growth of a socially relevant and dynamic service sector that can provide skilled jobs in villages to the young men and women who belong to families that do not have any productive assets.

MEETING THE CHALLENGES

The following are some of the principles that should form the basic guidelines for the agricultural strategy for the 1990s:

1. Water, energy, and cost-saving crop husbandry, i.e., efficient use of water and energy become the major pathways for enhanced production;
2. Grain-saving animal husbandry, i.e., the avoidance of animal husbandry practices that need large quantities of cereal grains to feed the animals; also, the establishment of rural biomass refineries;
3. To the factory but not to the town, i.e., the generation of adequate numbers of skilled jobs in the secondary agroprocessing and rural industries sector and the tertiary service sector so that those without assets or jobs need not migrate to towns and cities;
4. Sustainable management of the basic agricultural assets of land, water, flora, fauna, and the atmosphere; for this purpose current patterns of productivity measurement should be changed;
5. Income and employment orientation to agricultural research and development so that agriculture becomes not only an instrument for the production of more food grains and other commodities but also a catalyst for the production of more jobs and income in rural areas.

STRATEGIES

It would be useful to divide each African nation into the following science and technology zones; the precise technology applied in each zone should be tailored to the prevailing agroclimatic conditions.

Green Revolution Areas

These areas occur in most nations and are characterized by substantial improvements in crop productivity. A major challenge in these ares is to defend the progress already made and increase the yield further. This will call for the introduction of frontier technologies including the development of crop varieties that can give higher yields per cubic meter of water. Postharvest technology will also need additional attention so that all links in the producer-consumer chain are attended to in an integrated manner. Through cereal–legume crop rotations, animal husbandry programs can be supported at a higher level of efficiency than that of stall-fed animal rearing.

"Green-But-No-Green-Revolution" Areas

Several parts of Africa where water is not the most limiting factor fall under this category. Availability of water is not the primary constraint and the land is green most of the year; underground water has not yet been tapped extensively. The scientific management of water and nutrients holds the key to improving productivity in these areas and new methods of feeding the crops are necessary.

Dry-Land Farming

Saving and sharing water, new approaches to breeding and feeding crops, agroforestry systems with particular reference to leguminous shrubs and trees, animal husbandry, and off-farm employment—all these aspects will have to receive much greater attention. Steps to tap groundwater on a sustainable basis and the development of alternative cropping strategies based on different patterns of rainfall and moisture availability are needed. Such computer-simulation models should be supported by appropriate seed reserves of alternative crops. Just as grain reserves are essential for food security, seed reserves are essential for crop security.

Arid Ecosystems

The desert areas of the Sahel provide unique opportunities for high-quality seed production, arid horticulture, and animal husbandry.

Coastal Area Planning

The coastal areas afford opportunities for integrated agriculture and aquaculture production systems. Coastal agriculture can also be given an export orientation because transport costs can be minimized. For the purpose of planning coastal agricultural systems, 20 kilometers of sea surface and 20 kilometers of land surface can be considered an integrated entity.

Wetlands

Africa has a rich endowment of wetlands. Detailed planning is necessary for the sustainable management of wetlands in order to produce a variety of aquatic products and to avoid biological impoverishment.

Wastelands

The rehabilitation of degraded lands has to be undertaken using technologies most appropriate to each situation. Every effort should be made to restore wastelands to their full biological potential by the year 2000. A simple method of measuring the biological potential of land will have to be developed and popularized.

WATER: A KEY DETERMINANT OF THE AGRICULTURAL FUTURE

The global use of water doubled between 1940 and 1980 and is expected to double again by 2000, with two thirds of the projected use going to agriculture. Yet 80 countries with 40% of the world's population already suffer serious water shortage (WCED, 1987) and the competition for water for irrigation, industry, and domestic use will continue to increase. River water disputes are becoming common between nations and between states within a nation. Compounding these problems is the increasing pollution of water from a wide variety of causes—industrial effluents, agricultural activities, and chemical and human wastes. The conservation and scientific use of water must occupy a very high priority in any national agenda for sustainable development. I would like to suggest nine areas that need urgent attention.

Irrigation Extension

Despite a vast pool of information available on water management, the application of this knowledge at the field level is poor. Group endeavor on the part of the watershed, command area, or catchment area community is essential for efficient water use. We need an efficient extension system that will promote both individual and group action for the economic and effective use of the available water.

Area-Specific Water Management

Poor drainage is a problem in the irrigated areas of several African nations. Improved drainage and use of groundwater could increase both yield and the infiltration capacity of the soil. On-farm water and energy management is the key to improving production in semiarid areas but the precise priorities in water management will vary from region to region.

Farming-System-Specific Water Management

The water management system will have to be tailored to each farming system and the optimal use of available water should be a major aim of any farming system. Farming systems may involve multiple cropping, crop–livestock integration, or crop–fish combined production systems. The productivity of a farming system can be improved if the soil–water–plant relationships are studied in an integrated manner.

Aquifer Management

It is high time that every African nation develops a scientific groundwater management strategy with the cooperation of the farming community. This should ensure that we do not live on our capital. In chronically drought-prone areas groundwater sanctuaries should be established along the lines of those proposed in India during the drought of 1979.

Problem Soils

Water management in alkaline, saline, and heavy clay soil needs greater attention in relation to the farming system in use.

Problem Water

Sewage water, effluents from industries, and other types of water that normally cause pollution can be utilized in fisheries, for vegetable production, and for other purposes after it has been carefully treated and handled.

Management of Irrigation Lands

Unless the management of command areas of irrigations projects and watersheds becomes a joint-sector activity that is based on the shared goals and perceptions of farmers and government officers, inefficient and often scientifically harmful methods of water use will continue.

Operational Research Projects on Water Management

Operational research projects involving irrigation departments, appropriate universities and institutes, and farmers should be designed to demonstrate the value of water cooperatives in improving farm income.

Generating Consciousness of Water-Use Efficiency

In many parts of Africa, cereal productivity per unit of water has been found to exceed 1 kilogram per cubic meter (kg/m) under efficient water management. However, in practice it is usually less than 0.4 kg/m. Agronomists should express yield not only in terms of kilograms per but also in terms of kilograms per cubic meter of water. Farmers and administrators must also be sensitized to the potential adverse effects of improper water use. Salinization, waterlogging, and the spread of water-borne diseases can be avoided if preventive steps are taken.

ECOLOGICAL AND FOOD SECURITY AND BIOTECHNOLOGY

Ecological security implies the conservation and sustainable management of the basic life support systems of land, water, flora and fauna, and the atmosphere. It involves concurrent and integrated attention to all the components of the biosphere and geosphere. Ecological security is the only foundation on which enduring edifices of food and livelihood security can be built. The ecological security of our planet is under threat today because of human practices and the patterns of agricultural, industrial, and economic development and urbanization. Population growth is exceeding the capacity of natural ecosystems to remain ecologically sustainable in several parts of Africa. Increased emissions of carbon dioxide and other "greenhouse" gases arising from various human activities, particularly in developed countries, are leading to the phenomenon that is referred to as global warming. Deforestation of tropical forests is leading to a loss of biological diversity. These threats to the quality of life on each have led to a more widespread awareness of the fact that nature's patience is not inexhaustible. A disregard for the long-term consequences of current patterns of development as well as of population growth are leading to

1. air and water pollution;
2. soil erosion and land degradation;
3. deforestation and changes in micro-climate;
4. lost of biological diversity leading to enhanced vulnerability to pests, pathogens, and abiotic stresses such as salinity and alkalinity;
5. migration of the rural poor ("environmental refugees") to towns and cities as a result of loss of livelihoods in native villages;
6. global warming, acid rain, and increased ultraviolet radiation.

To reverse these trends, the World Commission on Environment and Development chaired by Mrs. Gro Harlem Brundtland, Prime Minister of Norway, has in its report titled *Our Common Future* (1987) urged all nations to accord the highest priority to making development ecologically sustainable. Although it is important to think about the distant future, it is even more important to plan for a better immediate future. The disparities between the lifestyles of the poor and the rich and in the amenities available in rural and urban areas are ever widening. We should examine the opportunities opened up by new technologies in this context.

Biotechnology Applications

It is becoming apparent that the new tools of molecular biology, genetic engineering, cell and tissue culture, microbiology, and chemical engineering can help to add a dimension of resource neutrality to scale neutrality in technology development. We can now plan to increase yield while reducing the cost of production. The tools of biotechnology and bioprocessing will particularly help to enhance the efficiency and economics of small-farm agriculture. If appropriate safeguards are strictly enforced the environmental hazards associated with recombinant DNA technologies can be avoided. If properly utilized, biotechnology can help to promote crop husbandry that can save land, water, energy, and money, and animal husbandry that can save grain. This in turn will decrease the amount of forest land being converted into crop land. Ecological security can be enhanced by an integrated application of market-purchased chemical inputs and farm-grown organic inputs. Considering that the population of most African countries is predominantly young, the integration of frontier technologies with traditional technologies will help to make farming both intellectually satisfying and economically rewarding; these are the necessary preconditions for attracting and retaining educated youth in rural professions.

ORIENTING STRATEGIES TO SPECIFIC TARGET GROUPS

Sustainable Development System for Small Farmers

The concept of a small farm varies from country to country. In India, a small farm is one which is below 2 hectors in size. Farms below 1 hector are classified as marginal. The cultivable land per capita declined in India from 0.48 hectors in 1951 to 0.26 hectors in 1981. It is expected to be 0.15 hectors in 2000. In China the per capita land availability is 0.11 hectors. In the semiarid, rainfed areas of Africa small farmers may often be cultivating 10 to 20 hectors of land but the yields are low and uncertain. The FAO's study of agriculture in the year 2000 shows that in most developing countries the average size of an operational holding will continue to decline. Besides the growth in population, the other factors contributing to this trend are the withdrawal of farm land for industrial, communication, housing, and other uses, and inadequate opportunities for employment in the nonfarm sector. Between 60 and 90 percent of the population in most developing countries lives in rural areas; for them, land- and water-based occupations consisting of crop and animal husbandry, fisheries, and forestry constitute both a way of life and a means of livelihood security. If current global trends continue, we will witness at the end of this century two contrasting systems of agriculture. In one system farmers will disappear and will be replaced by large agribusiness companies; in the other millions of families operating small holdings will be struggling to improve the productivity, profitability, stability, and sustainability of the farming systems that determine both the quality of their life and the dependability of the food security system of their countries. How can they be helped?

Before specific methods of helping small farmers are discussed, it would be useful to define the concept of sustainability.

Measurement of Sustainability

The core of the concept of sustainability lies in equal and concurrent attention to intragenerational and intergenerational equity. Intragenerational equity, which aims at giving a fair deal to the socially and economically disadvantaged groups, involves issues such as adding a dimension of resource neutrality to scale neutrality in technology development and transfer, thereby helping all farmers irrespective of the size of their holdings, their innate input-mobilizing, and their risk-taking capacity to derive economic benefit from new technologies. Intergenerational equity involves the protection of the productive base of agriculture. This implies the use of basic agricultural assets of land, water, flora and fauna, and the atmosphere in such a manner that today's progress does not ruin tomorrow's prospects. Thus, sustainability has both economic and ecological dimensions. What are its operational implications?

The first requirement is a new definition of productivity. At a meeting organized at the international Rice Research Institute in December 1987, the following definition was agreed to for the purpose of integrating economic and ecological sustainability in one measurable equation.

$$\textit{Productivity} = \frac{\text{Output value}}{\text{Input value}} + \text{Changes in environmental capital stocks}$$

$$\textit{Sustainability} = \text{Stability of productivity over time}$$

The value of output and input can be calculated with precision. We do not yet have readily usable and reasonably accurate methods of measuring the impact on environmental capital stocks. At least three areas need measurement tools.

1. Soil health. What is the impact of a given level of productivity on the biological potential of land? Soil health is the summation of the physical, chemical, microbiological, and erodability characteristics of the soil. Can we develop a reliable, speedy, and low-cost method of measuring the biological potential of land?
2. Water quality and availability. We need suitable criteria for measuring the impact of irrigation and crop intensification on the water resources of an area, such as the rate of exhaustion of the aquifer, increase in salinity, and so on.
3. Biological diversity. What is the impact of new technology on genetic erosion both in crops and farm animals? Can we develop methods of giving expressions to the impact of the spread of new technologies on biological diversity?

The standardized and low-cost methods of measuring and expressing productivity and its impact on environmental assets is urgently needed. This area of methodology development merits adequate technical and financial support. The emphasis should, however, be on simple tools of measurement that farmers can readily understand and adopt. Measuring sustainability by estimating the stability of productivity over time is best done by taking an entire village, watershed, command area of an irrigation project, or catchment area of a river system as the spatial unit for measurement.

Support Services: Package of Technology

To promote agricultural production on an economically and ecologically sustainable basis under conditions of small-farm holdings, there is need for reorientation of agricultural research in several areas. For example, yield is the product of interaction between the genetic efficiency of the crop plant or farm animal and the management efficiency of the farm family. The major management areas for yield enhancement that need concurrent attention are land and water management, crop and/or farm animal management, energy management, postharvest management, and farming systems management.

Scientific land and water management holds the key to sustainable agriculture. As mentioned earlier, we need simple methods of measuring the biological potential of land. Agronomic and agricultural engineering research will have to help the watershed community to upgrade degraded lands through appropriate land management systems including agroforestry systems. Social engineering research will have to devise public policy packages that can foster cooperation among members of a watershed/command area community in operations such as soil conservation, soil health care, and water harvesting and sharing. There is inadequate attention now on the social engineering aspects of soil health management.

Particular attention will need to be paid to breeding crops that can optimize yields under a given moisture regime and to feeding plants so that the genetic potential of new strains can be fully realized. Breeding strategies in particular need reorientation; we need to initiate a selection strategy where the F2, F3, and subsequent generations are grown alternately under conditions of moisture sufficiency and deficiency. Such selections under divergent conditions of moisture stress will help to breed varieties with the capacity to give high yields when moisture is adequate and reasonable yields when drought conditions prevail. The crop-feeding strategy should involve the use of legumes in the rotation (including stem-nodulating green manure crops), organic recycling, and balanced fertilizer application in accordance with the requirements for macronutrients and micronutrients in different soil types. The discovery of stem-nodulating *Sesbania* species such as *S. rostrata* in Senegal has opened up new opportunities for green manuring. A third aspect of crop management relates to the management of weeds, pests, and pathogens. In addition, cultural and cropping practices including date/density and manner of sowing and growing two or more crops together (in such a manner that the companion crops extract moisture and nutrients from different depths of the rhizosphere) will help to reduce and distribute risks and enhance yield and income. Risk aversion and distribution should be important goals of agronomic research, particularly in dryland areas, since often small farmers do not have access to protective irrigation.

Energy management relates to the energy needed (a) for irrigation, land preparation, and crop management including sowing and harvesting; (b) pest management; and (c) nutrient supply. In irrigated areas energy management, particularly for pumping ground water or distributing conserved water, is the key to the success of farming in years of drought.

Besides the handling of the part that is of direct human value (such as grain), efficient postharvest management implies detailed attention to every component of the biomass with particular stress on the preparation of value-added products from each of the components. Recent advances in bioprocessing, microbiological enrichment of cellulosic material, and chemical engineering have opened up new opportunities in biomass utilization.

In the ultimate analysis, packaging the best available technologies into an economically and ecologically desirable farming system should be the endpoint of all scientific efforts. For this purpose it is necessary to promote the optimal use of the climatic, land, water, labor, and credit resources available to the family. To achieve this goal, a two-pronged strategy is needed.

First, efforts are needed to achieve appropriate blends of traditional and frontier technologies. The frontier technologies that merit particular attention are biotechnology, including both molecular biology and bioprocessing, satellite imagery, computer technology, and microelectronics. Second, the design of the farming system—whether it is wholly based on crops, crops and animals, or crops, animal husbandry, fisheries, and forestry—should give explicit consideration to promoting better linkages among the primary (agricultural), secondary (industrial), and tertiary (services) sectors of economic activity in rural areas. Farming system research specifically designed to promote better growth linkages is necessary if we are to add the dimensions of income and employment generation in projects that are designed to increase food production.

Services such as knowledge and skill transfer, credit supply, and the timely availability of the inputs needed for higher productivity have been receiving attention in African countries. However, more research is needed on delivery systems, with particular reference to the following:

1. Synchronization in space and time between knowledge transfer and the inputs needed to apply that knowledge.
2. Synchronization in space and time between credit supply and the inputs for which the credit is intended.
3. Organization of the services sector in the villages, e.g., seed production and distribution, integrated pest management, and improved postharvest operations, particularly by rural women and youth. For this purpose, it will be necessary to train women in seed technology, integrated pest management, and improved postharvest operation and to help them to organize effective service organizations.
4. Helping to organize producer-oriented marketing arrangements and assisting farmers to avoid distress sales.
5. Promoting through training and appropriate infrastructure support the growth of industries designed to produce value-added products from every part of the agricultural biomass available in a village or in a group of adjoining villages. In general, research on delivery systems that enable small farmers to derive full economic benefit from new technologies has not received the attention it deserves. Such research should aim toward

drawing lessons from successes and failures in the field of technology transfer and helping to design self-replicating systems of technology diffusion. Similarly, training procedures should be largely based on learning by doing, particularly where illiteracy is still widespread. Adequate attention to software-development for the use of mass media such as television and radio is also essential if the investment in the hardware components of such media is to confer the anticipated benefit.

The public policy package should include incentives for group cooperation in areas such as land and water management, integrated pest management, integrated nutrient supply, improved postharvest technology, and organization of local-level associations for sustainable development. In addition to reorienting extension services on a community basis, suitable incentives such as group insurance, group credit, and group awards for community endeavor in the conservation of environmental assets will be helpful.

Besides public policy instruments for promoting community endeavor and participation in environmental management and in the adoption of knowledge-intensive and ecologically desirable technologies, an important but often neglected area of policy is the promotion of activities in areas where ecological linkages over widely separated areas determine a common agricultural future. This is particularly true in the Sahelian region with regard to measures for combating desertification. Public policy should also foster desirable research efforts. As emphasized earlier, an area of research relevant to sustainable agriculture is the integration of traditional and frontier technologies. With the powerful tools available today, agricultural technology can help to integrate scale and resource neutralities in technology development and transfer and help to generate skilled jobs in the rural sector. These tools can also help to generate technologies for women that can help to reduce drudgery and introduce the twin concepts of flexi-time and flexi-location.

Some emerging technologies that need particular attention from the point of view of integration with traditional technologies are molecular genetics and recombinant DNA technology, tissue and cell culture, sero-diagnostic techniques and vaccines for animal diseases, bioprocessing and preparation of value-added products from agricultural biomass, microbiological upgrading of cellulosic material, satellite imagery and remote sensing, computer technology, and micro-electronics. In this context, a distinction should be made between science and technology. Scientific research leading to new knowledge and material can be generated in research institutions and universities. Technology development is, however, best done in farmers' fields jointly by scientists and farmers. The failure to make this distinction often leads to difficulties in the diffusion of new methods and approaches because of economic and/or ecological factors. Technology development should become a joint-sector activity between scientists and farmers. We urgently need centers for technology blending and dissemination that are inspired by the goal of bringing the best in modern science to the service of small-farm families. In this task, nongovernmental organizations can play a key role.

Women

Equity involves paying adequate attention to the needs of the economically, ecologically, and socially disadvantaged sections of the population. Achieving growth with equity demands concurrent attention to economics, ecology, gender, and generations. In terms of gender, the emphasis has to be on enhancing the productive role of women and reducing drudgery in the household and in productive aspects of their lives. Various in-depth studies have been undertaken in Africa to identify and articulate the real issues confronting women in their triple role. The studies carried out so far in the field of agriculture indicate that in spite of the key role of women in crop husbandry, animal husbandry, fisheries, forestry, and postharvest technology, those in charge of formulating packages of technologies, services, and public policies for rural areas have often tended to ignore the productive role of women. Consequently, the development of technologies specifically tailored to women's occupations and the involvement of women in technology development and transfer have received inadequate attention from both scientific and administrative departments of governments. The same is true in the case of training. The very term "Home Science College" used in universities indicates the emphasis placed on the household responsibilities of women.

The priority areas that need urgent and integrated attention are

1. the hardships arising from unemployment, underemployment, uncertain employment, and unpaid or underpaid employment;
2. the drudgery that results from the predominantly unskilled nature of the work assigned;
3. the lack of flexibility in terms of working hours, place of work, and duration of work;
4. the adverse impact of labor-displacing technologies;
5. the lack of adequate maternity and health care, day care, and associated needs;
6. little or no access to credit and other production inputs;
7. inadequate opportunities for the acquisition of new skills and for extension assistance—this problem is compounded by the handicap of illiteracy;
8. discriminatory land or other asset ownership rights and social status;
9. paucity of organizations that can provide women with needed management expertise and collective strength.

The response to these problems has to be in stages. First, there are many things that can be done immediately with the technologies and infrastructure already available; second, we should initiate specific research and development programs that within the next 5 to 10 years can help to convert unskilled jobs into skilled ones, diversify opportunities for employment, and introduce flexibility in the place, time, and duration of work. Some of these guidelines can be implemented immediately. For example, in the short term we should build on the historic and current professional skills of rural women and enhance their beneficial impact on income and quality of life by removing the major handicaps and constraints. Since the dawn of agriculture the women of Africa have played a major role in the following areas: food crop production, seed production, including the management and conservation of biological diversity, renewable energy management, crop/livestock/fish management, and postharvest management, including the utilization of agricultural byproducts. What steps can our research and development institutions take to strengthen the role of women in these areas?

1. Seed production and management and conservation of biological diversity: Women traditionally have been seed selectors and preservers. Whether literate or illiterate, they have the ability to spot the healthy plants whose seeds they carefully preserve for sowing next year. The skills of rural women to produce hybrid seeds of high purity should be fully utilized. Similarly, in the area of conservation of biological diversity, women traditionally have been active conservationists. Their involvement in the organization and conservation of heritage sites, biosphere reserves, national parks, and community gene banks will enhance the effectiveness of the conservation movement.
2. Renewable energy management: Women traditionally have not only collected fuel for cooking but have also found nutrition for both crops and farm animals. Organic recycling and collection of fodder and feed have been their traditional occupations. Today there are many opportunities in the field of renewable energy management and utilization that women can handle effectively, if assisted to do so.
3. Crop/animal/fish/forest management: Experience in the Philippines and elsewhere has shown that women can organize integrated pest management involving biological and cultural control, genetic resistance, and need-based application of chemical pesticides in a very effective manner. This will help to prevent pest resurgence and epidemics. Similarly, women can handle biofertilizer production programs and help to organize and implement integrated nutrient supply systems. In the case of horticultural crops, the spread of integrated pest management and integrated nutrient supply techniques becomes even more important. The role of women in animal husbandry and aquaculture can also be further strengthened through appropriate training.
4. Postharvest management: Considerable improvement is possible in all the links in the producer–consumer chain. Preparation of value-added prod-

ucts is now feasible from every part of the plant and animal biomass. To achieve this, there are several ground rules that agricultural research and training institutions should integrate into all their programs. For example, job destruction and job creation must be concurrent events. The aim of technology should be labor diversification, drudgery reduction, and better energy conversion, and not labor displacement. Knowledge and skill transfer mechanisms should be tailored to the sociocultural and socioeconomic conditions of rural women and to the opportunities for remunerative marketing. Every effort should be made to increase the active participation of women both in technology development and dissemination. Also, knowledge and skill transfer should be synchronized and the supply of the necessary inputs, including credit; for this purpose, the growth of women's organizations and the strengthening of their professional management capacities should be promoted.

NEW FRONTIERS

Developments in several fields have opened up new opportunities for making rural jobs knowledge intensive and skilled and for introducing the dimension of flexibility in the place and duration of work. These are

1. Biotechnology, including bioprocessing, microbiology, tissue culture, and genetic engineering, and their applications in animal breeding and disease control;
2. Information technology, including computer-aided instruction, and extension and mobilization of mass media (particularly television and radio) to generate awareness and disseminate knowledge;
3. Space technology, including conservation monitoring and supporting ground floor work;
4. Micro-electronics, which lends itself to decentralized production;
5. Management technology, which would help in integrating all the elements of a technological package in a mutually supportive manner. In particular, the managerial and organizational aspects of decentralized production supported by a few key centralized services need attention. Also, the planned strengthening of the beneficial linkages among the primary, secondary, and tertiary sectors will have to be achieved.

In order to achieve the above, the following three steps should be taken immediately.

1. Technology blending centers should be organized in each agroecological zone in order to combine the good features of traditional and frontier technologies. Such centers should also aim to upgrade the productive role

of women. Focusing on the needs of women helps both men and women, whereas the reverse is not always true, as past experience shows. For example, demonstrations in small farmers' fields help big farmers also, whereas the reverse is not always true.

2. A dynamic program for training rural women in skilled employment should be launched. The training methods used should be adjusted to the level of literacy and the domestic obligations of the trainees.
3. Women's organizations should be fostered because they can serve as the link between the laboratory and the land.

AGENDA FOR ACTION

Early in 1980, the late Prime Minister of India, Mrs. Indira Gandhi, asked me to act as deputy chairman of the Planning Commission of India and help to develop the country's sixth five-year plan. When asked what her principal priorities in resource allocation were, she answered, "Work, water, and power." This summarizes beautifully the priorities for many nations in Africa also. I would therefore like to deal with them briefly.

Work

As emphasized earlier, greater opportunities for skilled employment are absolutely essential both for enhancing the purchasing power of poorer households and for attracting and retaining youth in rural professions. A massive effort in the blending of traditional and frontier technologies is essential for overcoming the death of work. Technologies that respond to these challenges must be knowledge-intensive, replacing unskilled jobs with skilled ones. Such technologies must lead to the diversification of labor rather than to its displacement. In addition, they must be ecologically sound. For this purpose, there has to be appropriate support in the form of educational and training opportunities. Pedagogical techniques should be tailored to the needs of each sociocultural group and women will need particular attention.

Water

Scientific water management is the key to elevating and stabilizing agricultural production. Fortunately, many countries in Africa have considerable untapped supplies of goundwater. To avoid the unsustainable exploitation of the aquifer, every country should develop a policy for its scientific and sustainable management. The management of water and soil has to be viewed in an integrated manner, otherwise the process of desertification will worsen. In chronically drought-prone and arid areas, drinking water sanctuaries will have to be developed for meeting human and livestock needs.

Power

Energy is often a major limiting factor in our efforts to increase productivity and improve postharvest handling and processing. Energy is also needed for pumping underground water and for a variety of other tasks. Energy requirements in agriculture include the energy supplied in the form of fertilizers, irrigation water, pesticides, and implements. Agriculture is the single most important solar energy–harvesting enterprise in the world. The benefits of solar energy can be maximized by producing a beneficial interaction between solar and cultural energy. An integrated energy supply system involving possible combinations of renewable and nonrenewable energy will have to be developed in each nation.

A carefully designed strategy for generating the needed amount of work, water, and power will help in the development of sustainable food and nutrition security systems. To this end, an agricultural system that balances the needs for efficiency, equity, and ecological sustainability must be developed. Such a system will also need mutually reinforcing packages of technology, services, and public policies. Every nation should establish a national strategy for sustainable agriculture. Any top-level coordinating committee established for this purpose should include representatives from farmers, women's organizations, and mass media.

CONCLUSION

In the ultimate analysis hunger can be made a problem of the past only by increasing production, enlarging the purchasing power of the rural and urban poor, and designing socially relevant intervention programs to look after the needs of preschool children, pregnant and nursing mothers, and the old and infirm. Attention to a safe and adequate supply of drinking water is as important as attention to the supply of calories, proteins, vitamins, and minerals.

The single most important requirement for achieving sustainable nutrition security is political commitment and action. Political will is usually expressed in speeches but often it is not converted into political action. There is no scientific or economic excuse for hunger today. Wherever hunger persists the political system must seek the reasons and apply the remedies.

I can cite many examples of political action that can help to reverse current trends in food production. If today the most important food security challenges in Africa are physical and economic access to food, ecological access to food will be the most important food security challenge in the twenty-first century. Building the ecological infrastructure essential for sustained agricultural advance is therefore a priority. This will call for accelerated efforts in tree planting and soil and water conservation in the year during which there is adequate soil moisture. Unfortunately, during years of adequate rainfall the resources available for programs such as "Food for Work" are greatly reduced. We should therefore design a "good weather code" and a long-term plan of employment guarantee for ecological security; the employment offered should include skilled and unskilled jobs. Second, an essential component of the national strategy for sustainable agriculture should be a careful and integrated deployment of the available technical, commodity, and financial resources. These resources, whether national, bilateral, or multilateral in origin, should be used in a mutually reinforcing manner. Finally, we need to develop innovative approaches for using the know-how and experience of educated women. Many of them are unable, for a variety of reasons, to take up full-time jobs. It is important that methods are devised to establish para-extension cadre consisting of women who are willing to teach rural women new skills on a part-time basis. Harnessing the skills of professionally trained women is an urgent necessity for enhancing the production of agriculture. Too much training is being wasted and while the brain drain should be minimized, the great challenge before political leaders is how to nurture the brains already in the country. The African continent is a land of opportunity. It is for the leaders and people of this continent to exploit the available opportunities to promote food, work, and health for all.

REFERENCES

Africa Leadership Forum. *The Challenges of Leadership in African Development* (Nigeria, 1988), 46 pp.

Africa Prize for Leadership 1989; Presidential Address, Zimbabwe.

Africa Recovery–new feature, 1989; The African Alternative: policy recommendations.

Babangida and Obasanjo, O., *Africa in Today's World and the Challenges of Leadership* (Nigeria: Africa Leadership Forum, 1988), 51 pp.

Dreze, J., *Famine Prevention in Africa: The Development Economics Research Programme, Suntory* (Toyota International Center for Economics and Related Disciplines, London), 67 pp.

Eicher, C. K., "Transforming African Agriculture," The Hunger Project paper, 1986, No. 4, 35 pp.

Food 2000: Global Policies for Sustainable Agriculture (London: Zed Books, 1987), 133 pp.

Food and Agriculture Organization of the United Nations (FAO), "Potential Population-Supporting Capacities of Lands in the Developing World," 1982.

Gritzner, J. A., "The West African Sahel: Human Agency and Environmental Change," The University of Chicago, Geography Research Paper No. 226, 170 pp.

Lewis, J. F., et al, "Strengthening the Poor; What Have We Learned?" U.S. Third World Policy Perspectives No. 10 (Washington, D.C., 1988), Overseas Development Council, 239 pp.

Myers, N., "Population Growth, Environmental Decline and Security Issues in Sub-Saharan Africa," Paper for SIDA Conference on Ecological Stress and Political Conflict in Africa (Stockholm, 1988), 21 pp.

Myers, N., "Mass Extinctions: What Can the Past Tell Us About the Present and Future?" Paper for IUGS/ICCF Workshop on Global Changes (Switzerland, 1989), 17 pp.

Obasanjo, O., "Issues in Agricultural Development in Africa: The Hunger Project," Tanco Memorial Lecture (Philippines, 1989), 10 pp.

Obasanjo, O., "Africa Leadership for Tomorrow's Continent," Interaction Annual Forum (Massachusetts, 1989), 5 pp.

Organization of African Unity (OAS), "Lagos Plan of Action for the Implementation of the Monrovia Strategy for the Economic Development of Africa," 1980.

Report of the Committee on African Development and Strategies 1985; *Compact for African Development*, eds. Lawrence S. Eagleburger and Donald F.MC Henry, 24 pp.

Rukuni, M., and Eicher, C. K., "The Food Security Equation in South Africa" (Manuscript, 1989).

Sen, Amartya, *Poverty and Famine: An Essay on Entitlement and Deprivation* (Oxford, U.K.: Clarendon Press, 1981), 257 pp.

Strong, M. F., "Ending Hunger through Sustainable Development," Tanco Memorial Lecture (Tokyo, Japan, 1989), The Hunger Project, 30 pp.

Swaminathan, M. S., "Building a National Food Security System." Aggrey-Fraser-Guigesberg Memorial Lecture of the University of Ghana (The Indian Environmental Society, New Delhi, 1981), 121 pp.

Swaminathan, M. S., "Sustainable Nutrition Security for Africa: Lessons from India," Tanco Memorial Lecture, 1986, Hunger Project Papers, No. 5, 35 pp.

Swaminathan, M. S., "The Emerging Global Agricultural Scenario," *Journal of the Royal Society of Arts*, 85 (1987): 891–911.

Swaminathan, M. S., Indira Priyadarshini Lecture, KRIBHCO, (New Delhi, 1988).

Technical Memorandum, U.S. Congress Office of Technology Assessment, "Africa Tomorrow: Issues in Technology, Agriculture and U.S. Foreign aid," 1984, 145 pp.

United Nations Economic Commission for Africa, "African Alternative Framework to Structural Adjustment Programmes for Socio-Economic Recovery and Transformation," 1989.

World Bank, "The Challenge of Hunger in Africa: A Call To Action," 1988, 15 pp.

World Commission on Environment and Development, "Our Common Future" (New York: Oxford University Press, 1987), 183 pp.

World Food Council, "The Cyprus Initiative Against Hunger in the World, President's Report to the 15th Ministerial Session (Cairo, Egypt, 1999).

Chapter 3

Food Security in Africa

United Nations' Economic Commission for Africa

INTRODUCTION

The Concept of Food Security

In the past the concept of food security was left undefined, perhaps by default, because it had defied definition. More recently, however, food security is generally accepted as entailing not only food availability through domestic production, storage, and/or trade, but also, and perhaps more importantly, food access through home production, purchase in the market, or food transfers.

The Committee on World Food Security has defined the three specific aims of food security (FAO, 1983) as

1. ensuring adequate production of food;
2. maximizing the stability of food supplies;
3. ensuring access to food, particularly for those in greatest need.

Underlying this new concept of food security are two assumptions, that until recently had not been given sufficient attention by policymakers in African countries. These assumptions are that a distinction must be made between permanent and temporary food insecurity and that famine, hunger, and malnutrition are, in most cases, as much problems of poverty as of shortage of food supplies.

Temporary food insecurity is the result of short-term fluctuations in production brought about by fluctuations in household incomes, food consumption, and availability of food at the national and global levels. Temporary food insecurity is thus a manifestation of temporary lack of access to sufficient food and can therefore be eliminated or drastically reduced by corrective stocking policies. Because enough food has traditionally been available at the global level, the problem of temporary food insecurity in the African context must be seen largely as one of inability to buy. The important variables are the magnitude of the instability of international prices and the average levels of real incomes prevailing in African economies. Permanent or structural food insecurity represents a state of continuous inadequate access to food and is caused by a long-term structural deficit in production and the chronic inadequacy of purchasing power.

A recent study by the ECA (1987) of ongoing food security programs in Africa concludes that the majority of international donor agencies that have assisted developing countries in the implementation of food security programs and projects in the past have been preoccupied mainly with the elimination of temporary food insecurity. National food security programs in Africa can be grouped under one of the following five headings:

1. Programs aimed at the improvement of food supplies in emergencies;
2. Measures to improve internal marketing and distribution networks and their functioning;
3. Measures to increase national preparedness;
4. Reduction of preharvest and postharvest food losses;
5. Advisory services for assistance to and improvement in food security planning and programming.

A listing of various subregional, national, regional, and international food security projects can be found in the Appendix to this chapter.

Food Security, Famine, and Poverty

It is now generally accepted that the problem of food insecurity is not just about food alone but also about the general problem of poverty and unequal distribution of purchasing power among and within regions and nations. Famine, for example, represents the most severe form of temporary food insecurity manifesting itself in a sudden collapse of food consumption and a dramatic increase in the incidence of diseases and death. Because famine is caused by a variety of factors including crop failures, and natural calamities such as wars, floods, and, sometimes, high food prices, it often occurs without substantial reduction in food availability per head or even in the face of food being exported from the affected country or region.

A review of past famines, food scarcities, and shortages would, if not reveal the degree of risk to which countries or areas are exposed, certainly pinpoint the areas of greatest risk. The last exhaustive survey of famines was made at the end of the nineteenth century. The oldest recorded famine dates from about 5000 years ago and information on it was carved on the Stele of Famine found at the first Nile cataract in Egypt. Ancient famines have been recorded in Egypt and Ethiopia; the latter is said to have experienced about 10 major famines each century. Famines are also known to be a recurrent feature in the arid and semiarid zones of Africa. Kenya is said to have experienced 15 famines and noteworthy food shortages between 1879 and 1979; these years served as reference points for many people interviewed during a census conducted in 1979. Over the last 35 years, famines have occurred in the Republic of Congo (immediately following independence), in Nigeria between 1967 and 1969, in the Sahelian zone between 1965 and 1974 (where several hundred thousand people died), and in Ethiopia in 1974 (costing lives and bringing about profound political changes) (Leibenstein, 1982). A considerable number of famines are recalled in Rwandan oral history. The 1984 famine in Ethiopia is too recent to need any further mention.

AN ASSESSMENT OF THE FOOD SECURITY SITUATION IN AFRICA

Periodic but regular continent-wide consumption surveys represent an ideal way of assessing the food security situation in Africa. However, taking into consideration both logistical and cost considerations, these surveys appear unfeasible at the present time so alternative ways of measuring the level of food insecurity in the continent must be used.

In the short term and the medium term, the measurement of food insecurity in Africa has to take into account the variability of staple food production, the degree of preparedness of countries to foresee and cope with a widespread and acute food shortage, and their ability to obtain access to commercial food grain stocks on the world market. All early attempts to improve food security have concentrated on stocking and import policies for food grains, principally cereals (predominantly wheat), and these grains have been the principal component of emergency aid. However, it must be recognized that for several reasons it is not appropriate to limit these foodstuffs to cereals or even to grains in general. The cereal portion of the staple diet in most African countries is not as important as it is in Asia, for example. In some countries in sub-Saharan Africa the cereal portion of the diet is so low as to make assessment of the food situation based on cereals alone entirely inadequate. For example, the proportion of cereals in total food consumption in Zaire for the period 1979–1981 was 14.5 percent. Any assessment of food insecurity in Africa must take into account roots, tubers, nuts, oilseeds, fruits (which includes plantains), and, in certain countries, even alcoholic beverages.

Although it might be claimed that food insecurity is properly measured by a consideration of consumption patterns and that consumption could be defined as national production plus net imports (taking into account any changes in stock level), in practice it is extremely difficult to obtain all the necessary data for a sufficiently recent time series to be constituted. Furthermore, data on carryover stocks are difficult to come by in most developing countries. Valdés and Konandreas' 1981 study of consumption instability in 24 developing countries from 1961 to 1976 has shown that there is a high positive coefficient of correlation between total staple food production and consumption. This is to be expected if in general imports are insufficient to cover production shortfalls. This is particularly true for sub-Saharan Africa where fluctuations in staple food consumption closely follow the fluctuations of domestic staple food production. Any assessment of the food security situation in Africa should therefore be informed by this consideration.

The most recent assessment of the food supply situation in sub-Saharan Africa (FAO, 1989) reveals that aggregate production of cereals in the 45 developing countries of sub-Saharan Africa was a record 58 million tons, some 30 percent higher than in 1987 and higher than the long-term trend. This was because of good harvests in most countries, including above-average crops in all Sahelian countries, with record outputs in many of them.

Notwithstanding this substantial improvement in cereal harvests in 1988, all 45 countries of sub-Saharan Africa will continue to import cereals, mostly wheat and rice. The FAO estimates that the aggregate cereal import requirements of these countries in 1988–1989 is 8.2 million tons, about 7 percent less than their actual imports of the previous year. The fall in the import requirements was most dramatic in Eastern Africa and, to a lesser extent, in Southern, Western, and Central Africa, respectively.

Over 50 percent of the aggregate import needs of these countries will come from commercial purchases. Most of the rest will be covered by food aid, with a large part of this going to the seven countries considered to be facing continued food emergencies—Angola, Ethiopia, Malawi, Mozambique, Sierra Leone, Somalia, and Swaziland.

The FAO assessment concludes that cereal prices will remain relatively high (in mid-April wheat and maize prices were 37 and 20 percent higher, respectively, than their 1988 levels) and that, although the region's import needs will be less than in 1988, several African countries will be seriously affected by the higher prices of most cereals in the international market. FAO's latest forecast is that the cereal import bill of the 45 countries of sub-Saharan Africa in 1989 will rise about 40 percent to about US$ 1000 million, even though the tonnage imported will be lower. The fact that these escalating import costs will come at a time when many of the African countries will be facing growing external debt burdens, higher interest rates, deteriorating foreign exchange reserves, and restrictions on imports as part of government austerity and structural adjustment programs, suggests that many African countries will not be able to acquire the cereal imports they require. This is bound to lead to further deterioration in the food security situation in the continent.

The longer-term prospects of food security in the continent are not any better. The FAO (1986) has estimated that if present trends continue, by the year 2010 food imports into Africa would cost US$ 28.5 thousand million at constant prices compared to agricultural export earnings of, at most, US$ 12 thousand million. The cereal gap alone would grow to 100 million tons, 58 million of which would be in sub-Saharan Africa. No plausible combination of commercial food imports and food aid could meet the deficits of the low-income countries of sub-Saharan Africa on such a scale. The FAO also concludes that a continuation of current trends would place many countries on the edge of survival even if they were provided with substantial international support. Although other food surplus countries could produce sufficient food to meet the deficits foreseen if present trends persist, the volume of food imports and food aid necessary would exceed the existing, and probably the foreseeable, transport and distribution facilities of many African countries.

The scenario described above suggests that the challenges of agricultural production and food security in Africa are indeed enormous and that African leadership will be called upon to design and implement some difficult but imaginative policy choices if the continent is to be emancipated from the throes of hunger, poverty, and famine. In later chapters we describe some of the challenges that would confront African leadership with regard to increased food and agricultural production, sustaining production through more effective resource management, and ensuring accessibility to food through better marketing and trade policies.

FOOD AND AGRICULTURAL PRODUCTION

With the exception of a few countries, the continent of Africa could be said to be facing one of the greatest economic development calamities of the century. The economic depression of the 1930s pales in significance when compared to the current economic crisis facing the continent.

Of the 39 countries listed by the World Bank as being the poorest in the world in 1988, 24 are in Africa (World Bank, 1988). One of the major reasons why many of these continue to remain poor is because they have been unable to cope with the food and agriculture demands of their rapidly increasing populations. During the current decade, agricultural production in Africa rose by less than 2 percent per annum and food production per capita was generally below what it was at the beginning of the decade. As population pressures have increased, the amount of arable land per capita in the continent has decreased, limiting fallow periods and the possibilities of increasing production through extensive cultivation as was done in the past.

Most African countries are also faced with the double-edged tragedy of either feeding their population or paying their debts. By way of inherited production and consumption structures, many of these countries have been conditioned to emphasize the production of raw materials while engaging in "jet-age consumption habits" that require the generation of new investments. These investments depend on the availability of financial resources in foreign currencies for the purchase of machinery and processed raw materials for which the continent does not have a comparative advantage. What is worse, as population pressures have increased, foodstuffs have joined the list of essential import items of the continent even though the prices of these items continue to increase and the money to buy them becomes harder to obtain.

Earlier we had adopted the broad definition of food security: that national food security is not simply synonymous with national food self-sufficiency but involves both availability and access to food. It is, nonetheless, very clear that an important element in the attainment of national food security as defined in this chapter is food self-sufficiency and agricultural development. A nation's poor and hungry people are unlikely to have access to enough food throughout the year to lead an active and healthy life unless they produce their own food in adequate amounts. Neither could their poverty and hunger be eliminated in the absence of widespread agricultural and rural development.

The Economic Commission for Africa (1989) believes that the structural characteristics identifiable with the pattern of production, consumption, and exchange of the African economy constitute the most fundamental causes of its underdevelopment and retrogression. These characteristics, of course, owe their origins to the colonial administrations of western European countries such as the United Kingdom, France, Belgium, the Netherlands, Germany, Portugal, and Spain, and are at the heart of Africa's contemporary food security crisis.

Most African economies are still largely agrarian in structure and outlook. Because food and agricultural production form the principal activity in these economies, both in terms of their contribution to the occupational distribution of the labor force and in terms of their proportionate contribution to the gross national product, the attainment of sustainable food security in Africa must be anchored in changing the present structure of production and employment so that more food is produced by the right people employing the agricultural technologies most appropriate to their needs and circumstances. By so doing, employment opportunities and the overall ability of the poor people of the country to buy food would be enhanced. This is a formidable food and agricultural production challenge for attaining food security in the continent, one that will involve the development of appropriate technologies through relevant agricultural research and better access to factors of production by all rural groups, particularly peasants.

Given the agrarian nature of African economies, the essential precondition for overall social and economic growth in the continent is a dynamized agricultural sector brought about by a steady increase in agricultural productivity resulting from technological change. For African countries to get out of the "food and debt" trap that they are now in, domestic food production must increase so as to feed the growing population and export production must rise to generate the necessary foreign exchange to pay off the continent's debt and meet essential import needs. Both these developments will require agricultural research to generate new technologies that will facilitate the improved levels of crops and livestock production needed to dynamize the agricultural sector.

Agricultural research can be simply defined as an organized investigation undertaken to discover new and improved methods of production that will result in increased productivity and efficiency. Agricultural research can be grouped basically into two types: basic and applied. Basic agricultural research seeks to develop new and fundamental knowledge that can be applied to solve real agricultural problems; applied research adapts the new technologies to solve practical farm problems.

A major FAO study (1985) has shown that despite its present predicament, the potential for increases in agricultural output growth in Africa is substantial even with the existing low input levels that prevail in the continent. Increases in production could also be achieved if the majority of the continent's farmers could be assisted to move to higher levels of input use. All available evidence suggest that yield performance for principal crops are far below the levels that have been achieved at the research centers (Abalu, 1987). Technical scientists have identified low plant populations, insufficient fertilizer use, poor control of weeds, and nonavailability of suitable varieties as the major technical reasons for the low productivity of African agriculture. Several attempts have been made to apply to Africa technologies that have successfully addressed these technical problems in developed countries. These transfers have, however, turned out to be largely unsuccessful because they have often been inappropriate for small farmers, who constitute the majority of the African population, produce the bulk of the continent's total agricultural production, and will continue to do so for the foreseeable future. In a few African countries the few farmers who have large parcels of land have managed to get good liquidity and some education so as to be able to adopt some of the available improved technologies; however, the vast majority of African farmers have not adopted the new technologies as quickly or as readily as these "progressive farmers." In the past the low adoption rates among the mostly small-scale farmers was erroneously attributed to innate psychological characteristics that could only be broken down through increased persuasion and demonstration.

Continued low rates of adoption, despite concerted efforts to convince African farmers of the benefits of available improved technologies, have now led many scientists to believe that there may be other reasons why these farmers may not be readily or successfully adopting these technologies. For example, in their surveys of the evolving technical and social conditions of African agriculture, Malton and Spencer (1984) conclude that, with the exception of relatively limited high-potential zones, the set of new technologies available to African farmers is often inappropriate, responds poorly to farmers' changing conditions, and cannot bring about a sustainable response in aggregate supply.

Improving this situation demands a reorientation of the objectives and methods of technical agricultural research. The type of agricultural research that is appropriate for any country will depend on the unique technical and socioeconomic conditions that operate there and the financial resources and personnel available. It will also depend on the amount of basic research that is already available, and on how much of it can be beneficially applied or adapted to given situations and under given circumstances. In this regard basic and applied agricultural research are complementary. What is important, however, is that every African country must evolve an appropriate national agricultural research system that is charged with the key responsibility of generating and adapting the technology required to meet the agricultural development challenge of the country in a coherent and systematic manner. The research system appropriate for each African country will need to center around the agricultural technology and transfer process based on farmers' needs and circumstances and be serviced by an extension system that responds quickly and effectively to farmers' needs and reactions. The system should serve farmers who can benefit from the available technologies and whose responses to extension and research workers are promptly taken into consideration in further research efforts.

The process of improving upon the agricultural technology in Africa will therefore involve many participants in a two-way process, one that begins at one end with agricultural scientists working at research stations and extends to include the farmers themselves. The range of activities involved in this process includes basic or fundamental investigations, research conducted at experiment stations, adaptive experiments on farmers' fields, and experimentation by farmers themselves. This range of activities must also be accompanied by a strong national extension system as well as other agricultural support services.

RESOURCE MANAGEMENT FOR SUSTAINABLE FOOD PRODUCTION

Increases in food and agricultural production per se will not guarantee food security for the peoples of Africa. The increases in food production must come from a sustainable food production system. Because African food production systems are generally resource constrained, organizationally complex, and ecologically vulnerable, their sustainability can only be ensured if they are managed effectively. Among the important factors that impinge on the sustainability of food production systems in Africa are environmental degradation (soil loss, deforestation, and desertification) and erratic weather conditions.

Soil: A Valuable Resource for Life Support Systems

Soil is one of the most valuable resources of the continent; it is one of the most important support systems on which other systems (forests, atmosphere, and sea) depend. In Africa about 8 percent of the soil is found on fertile lands whereas 92 percent is found on marginal lands (deserts, and acidic, saline-, and water-logged soils). The United Nations Environmental Programme (UNEP) has estimated that the area prone to desertification worldwide is approximately 30 million square kilometers, of which 6.9 million square kilometers (23 percent) lies in Africa south of the Sahara. The semiarid soils are usually saline and therefore not suitable for crop production.

Because soil is a vital resource, it must be conserved at all costs. Available data suggest that many African countries have lost and continue to lose most of their precious topsoil through land impoverishment (soil erosion, overgrazing, overcultivation, etc.). It has been estimated, for example, that Ethiopia loses 1 to 3 billion tons of fertile soil every year. Furthermore, water erosion poses great risks in countries such as Lesotho, Sierra Leone, Guinea, Ghana, Zaire, and Liberia, and wind erosion poses similar threats in Mauritania, Senegal, Niger, Sudan, and Somalia. There is a painful saying that Lesotho's only exports are soil and cheap labor. Soil loss is not only a national problem but also a transboundary issue. It has further been estimated that about 742 million hectares (more than one quarter of the whole continent) is in the process of becoming useless for cultivation.

During the colonial era in Africa, soil conservation was a matter of serious concern. However, the colonial governments did not emphasize the significance of land conservation to the African farmers; instead land conservation practices were rigidly imposed. The consequences were always severe for those who failed to observe the prescribed soil conservation measures, such as planting grass strips or making contour trenches; indeed, it was considered illegal to cultivate on river banks. After independence very few African governments vigorously pursued the legacy of their colonial masters. Despite the fact that soil loss is a serious threat in many African countries, there are very few or no successful stories to build on; instead there are many grim tales of land mismanagement that has led to desertification.

Soil conservation practices can only be implemented by the farmers themselves. It should therefore be the policy of the African governments to provide advice to individual peasants through a network of agricultural extension services. The extension workers must be able to recognize the type of soil erosion and determine other areas where erosion is also likely to occur. Armed with this information, they should then prescribe the appropriate methods of protecting the land from soil loss. The extension workers must, however, not lose sight of the existing traditional methods of soil conservation; these methods must be improved and promoted. For governments to execute these obligations, they must in the first instance develop sound national soil policies. Women, who do most of the farming in Africa, must feature prominently in the planning and implementation of agricultural productivity and soil conservation.

Rainfall Fluctuations in the Sahel: A Historical Perspective

The climate of a region or subregion cannot be assumed to be a constant phenomenon. During this century the rains in the Sahel were 20 percent below normal in 1903, 1909, 1912, 1917, 1937, 1943, 1969, and 1974. However, of greater importance is the fact that the rainfall patterns from 1900 to 1917 and from 1965 to 1985 were fairly similar. During these periods the rains in the Sahel were below normal.

For four decades (1902–1960) rainfall in the Sahel region was above normal. Indeed, underdevelopment and impoverishment, which according to the colonial masters characterized the tropics, were closely associated with the then-constant hot and humid tropical climate. Because of the constancy in climate, it was assumed that regions that received high rainfall would also have a correspondingly high agricultural production in contrast to those that received low rainfall. The Sahelian drought, which commenced in 1968 and continued into the mid-1980s, changed the thinking of those who subscribed to this view of the constancy of climate. Today climate is no longer assumed to be an unchanging phenomenon. The shifts in temperature and precipitation are important parameters for all societies and especially for those that depend on agriculture for their livelihood. Whether these climatic fluctuations in the Sahelian subregion are caused by increases in "greenhouse gases" remains to be established.

Between 1968 and 1973, 15 African countries experienced untold suffering from the ravages of drought. It is estimated that 150,000 people and millions of cattle and other livestock perished in this region as a result of drought and related causes. The 1982 to 1985 drought also resulted in the deaths of human beings and livestock, as well as causing the displacement of families and communities in 34 African countries. Today 36 countries are affected by drought: 5 states in the northern Sahara region, 21 in the Sudano-Sahelian region, and 10 in the Kalahari desert area. Nineteen of the 36 states fall into the category of least developed countries and about 200 million people live in these drought-stricken countries.

In Africa prolonged droughts such as the Sahel drought of 1968 to 1985 generally spread famine throughout the society. It is always the poor who experience famine first. In addition to directly affecting crop yields, drought triggers migration of pastoralists and nomads from rural areas to urban centers.

Because man is unable to prevent droughts, his next best option is to predict their arrival so that appropriate action can be taken to mitigate their adverse effects. Of particular interest and concern to African countries should be developing and/or strengthening institutional capability to mitigate the effects of drought. Efforts should also be made to encourage neighboring countries to cooperate in the exchange of information and experience on early warning systems and to develop remote sensing techniques that can predict droughts.

The Future Climate of Sub-Saharan Africa

Studies on climate change have assumed great significance within the last few years since the Villack meeting in Austria, 1985. The scientists attending the conference announced that, "As a result of the increasing concentrations of 'greenhouse gases,' it is now believed that in the first half of the next century a rise in global mean temperature could occur which is greater than any in man's history." They predicted a rise in global temperature of 1.5° to 4.5° C. Although there is a consensus on the nature of global climate change, the nature of climate change at the regional level is not yet well understood. In Africa and particularly in sub-Saharan Africa, climate is already exerting control over development programs. Future developments in Africa will have to be planned carefully bearing in mind the sensitivity of the region to climatic variations.

The so-called greenhouse gases that cause climatic changes include mainly carbon dioxide, oxides of nitrogen, methane, and chlorofluorocarbons. These gases allow the passage of incoming solar energy but they trap the outgoing radiation from the earth, resulting in the warming of the earth. In addition, these gases affect the earth's stratospheric ozone layers. If the emission into the atmosphere of greenhouse gases remains the same, the Villack conference predicted that climate change will occur in about 2030 to 2050. However, the warming will not be uniform globally.

An important question to ask is to what extent will sub-Saharan Africa be affected by climate change? In 1978, UNEP sponsored a meeting of experts who noted the present seasonal and annual climatic variations in the semiarid tropics as well as the decline in rainfall, which results in drought and desertification. It was predicted that future climatic change would deepen the crisis of food, water, and energy. Because of shifts in natural ecological zones (i.e., crop and forest belts moving to higher latitudes), the pressure on marginal and virgin ecosystems will increase. The experts also predicted that due to enhanced evapotranspiration, the humid tropics will become drier.

The sea level will also rise due to the change in climate (about 25–140 cm), which will increase flooding and salinization. Undoubtedly, the sea level rise will disrupt settlements, agriculture, industry, forests, fisheries, and wildlife habitats as well as increase salt water levels of groundwater, rivers, and bays in the humid tropics. The change in climate will, therefore, have a significant impact on the social and economic development of this region. For example, it is predicted that such a rise will affect about 20 percent of Egypt's 35,000 square km of arable land.

What can be done to avert the impending calamity of climate change? Of the greenhouse gases mentioned above, carbon dioxide is responsible for 66 percent of the global warming. The preindustrial concentration of atmospheric carbon dioxide was 280 ppm; in 1980, it was 340 ppm and by the year 2030 it is expected to rise to 560 ppm. Carbon dioxide in the atmosphere can be reduced by burning less fossil fuels such as coal and oil and, in addition, atmospheric carbon dioxide can be reduced through the process of photosynthesis.

Higher plants fix atmospheric carbon dioxide into sugars in their green leaves. In this fundamental biological process (photosynthesis), oxygen is released into the atmosphere. Forests, therefore, clean the air by taking up the carbon dioxide and releasing oxygen. In this process, forests regulate the world climate by absorbing the pollutant carbon dioxide. Deforestation results in an excess of carbon dioxide in the atmosphere, producing the greenhouse effect. It is estimated that in Africa, about 3.6 million hectares of forests are cleared each year. Cote d'Ivoire has the highest deforestation rate in Africa (70 percent). Burning of forests and grasslands, which is also common in sub-Saharan Africa, releases carbon dioxide into the atmosphere and massive reforestation programs are therefore needed. Setting aside one day a year for tree planting is not adequate—reforestation must be carried out throughout the year and the planted seedlings must be looked after keenly, which includes protecting them from goats and sheep.

FOOD ACCESSIBILITY

Access to food encompasses physical and economic aspects. Physical access to food relates both to the adequacy of supply and to the efficiency of the distribution system, including storage, preservation, transport, marketing, and processing. Economic access to food relates to the ability of groups of people to establish entitlement over a requisite amount of food.

The challenge of improving food security in Africa would, therefore, include actions aimed at improving the procurement system and the food marketing system as well as actions aimed at providing the people at risk of hunger with income that would permit them to purchase the food they require.

The procurement system can be improved either by boosting local production or by importing to supplement domestic production. Other ways of enhancing the procurement system include the drawing up of a multiyear food requirement plan, the establishment of food security stocks, the prevention of postharvest food losses, and the establishment of an early warning system.

Improving the distribution system requires the development of adequate infrastructures, such as markets, roads, transport, and storage; the establishment of market intelligence information; and the provision of incentives to both producers and traders. The incomes of households will also improve depending on government policies regarding employment and transfers. These three components necessary for improving food security are interlinked in that an improved marketing system could considerably improve the procurement system, which in turn would contribute to raising the incomes of the majority of the population. In view of the close relationship of these three components, the importance of the marketing system in food security policy is great because it can promote dependable supplies of low-cost food for rural and urban populations. The major problems inhibiting the development of an adequate marketing system include government policies regarding food trade, pricing, and storage.

General Government Policy and Food Marketing

In the majority of African countries, if not in all, governments have used direct or indirect means to influence the trade of agricultural commodities, including the marketing of locally consumed foodstuff, export crops, and import food items. Such an intervention is guided by social considerations including the following:

1. Ensuring that food is accessible to everyone, especially to inadequately nourished citizens;
2. Establishing a developmental strategy through the stabilization of agricultural commodity prices in order to transfer revenues from marketed agricultural surpluses to the secondary and tertiary sectors;
3. Adopting a political strategy in order to provide incentives to producers or to favor powerful urban consumers.

In order to achieve these objectives and effectively control the market, governments in most African countries have established commodity marketing boards and have promulgated laws regulating food trade (domestic food marketing, export of agricultural commodities, and import of foodstuff). In so doing, they also have been keenly aware of their responsibility in developing marketing services including the construction of roads and other communications facilities, assembling centers, wholesale stores, retail markets, and so on, and the establishment and enforcement of basic commercial legislation. Although it is normal that a government be involved in agricultural marketing with a view to achieving national economic development and ensuring improved food security for its citizens (particularly those of its low-income consumers), the experiences of African governments so far in this regard have not always been very encouraging.

At first, the publicly managed marketing boards were, at best, able to procure only about 50 percent of the total food sales. Also, many studies have reported the inefficiency of the public marketing bodies caused by mismanagement and serious lack of financial means. Besides, many governments have regarded the parastatals as a means of restricting the role of private traders and administrative and technical staff members have usually maintained an attitude of hostility toward the private entrepreneurs who are seen as opponents to the public marketing system. Also, the subsidies that governments constantly allocate to the marketing boards often bring about unfair competition with the private traders and, in many cases, lead to disincentives in the food marketing sector. Besides its direct involvement, governments can contribute positively or negatively to promoting efficient food marketing because they have the responsibility of developing marketing infrastructures.

There is no doubt that laudable efforts have been made by national governments to upgrade ports and railways and to construct roads and other communications facilities. Also, governments (assisted by municipalities) have established a sizable number of city markets. It is particularly noteworthy that African governments have invested in highway construction, making accessible, if not most of the production areas, at least the collection centers and increasing the volume and speed of traffic. These elements have contributed to improving the commerce system in general and to some extent the food procurement system. However, there is still much to be done in order to organize a large turnover because, as in the subsistence farming context, food production for the market is still irregular and scattered. Many public markets are badly maintained, unhygienic, and have poor waste disposal facilities. This is because the markets are usually designed taking into account only architectural elements and ignoring practical aspects.

Governmental Pricing Policies

The instrument of food pricing policy is very complex, particularly when related to food security. If the "spirit of enterprise" of private operators contributes to the effectiveness of the food marketing system, then there is no doubt that private marketers operate on the profit motive and tend to sell food to better-off consumers in order to secure higher profit margins. Also, it is very common for powerful traders to use various social and political privileges to control food markets. The laissez-faire attitude of governments can therefore lead to market imperfections and to artificial scarcity of food. Food prices are therefore of great concern to governments in their efforts to make foods accessible to all groups of consumers.

For locally consumed food commodities and especially for cereal crops, state intervention in price formation, administration, and management was initiated during the Second World War in Eastern and Southern Africa where many colonial farmers had settled; early in the 1960s such state intervention occurred in all other countries. The major objectives of pricing policy are sometimes in conflict with each other as governments attempt to provide incentives to producers and cheap food to nonproducers, particularly urban citizens. Price formation often involves several departments including the ministries of agriculture, commerce, and finance, and the state-controlled grain marketing boards. Also, the criteria used in determining price levels are either not clearly defined and/or are based on unreliable data about market situations. The fixed prices that are set relate to procurement price (producers' price) and to retail price (consumers' price) and are administered either through a panterritorial pricing system (common to all producers or consumers throughout the country) or zoning pricing systems. Although the dual-objective rationale for the system, which is to keep a level of incentive for producers and to ensure that food is affordable to almost everyone, is noble, some of its aspects require critical assessment; these include the unresolved conflict of interest between producers and consumers and restriction of competition.

Unresolved Conflict of Interest Between Producers and Consumers

This dilemma is embarrassing because it seeks to stimulate production and protect farmers' income and at the same time to keep consumer prices low enough so as not to affect the nutritional welfare of low-income urban people. Governments have a hard time deciding on whose interest is paramount. The fixed prices of the current pricing policies adopted in most African countries do not necessarily represent a sound appreciation of market forces and there is no clear determination to promote the producers' interest. According to FAO estimates based on data from 27 African countries, producer cereal prices and retail food prices for 1978 to 1980 increased by 14 and 10 percent respectively as compared to those for 1969 to 1971. Also, compared to other developing regions, the real producers' prices of cereal for African farmers rose by less.

Furthermore, even in cases where the level of producers' prices provides real incentives to boost production, price policy alone will not be enough to sustain the expected results. For a farmer who has no means of transportation to get the produce to market or has no market outlook, any incentive to grow more produce will quickly dwindle. Therefore incentives should not be simplistically limited to price level.

Restriction of Competition

If public parastatals are chosen to be the basic institutional framework for implementing and administering pricing policy, the dynamism of the private operators involved in the food marketing sector will be adversely affected. In view of the poor performance of the public institutions and the excellent performance of energetic women traders, particularly in West Africa, there is a serious need to consider restricting the public role to establishing the general policy for guiding private operators and for controlling the fairness of marketing operations. Also, pricing policies that tend to favor both producers and consumers by fixing procurement prices at satisfactory levels and reducing marketing margins will affect the whole marketing system. In such a situation, public institutions will build substantial deficits that will further absorb scarce resources and private operations will not invest in the marketing sector at such high risks. Therefore any pricing policy must also take into account the interest of the private food traders and, more particularly, the setting of price levels should take into account the costs to the marketing operator, particularly transport differentials. If the fact that the price elasticity of demand for food of the poor is high and can determine the fluctuations in per capita food consumption justifies that governments intervene in times of precipitous price rises, their permanent intervention in price formation could lead to supply shortages and subsequent food insecurity.

Storage as a Component of Food Security Policy

The constitution of food reserve at the household level is a component of subsistence farming and the millions of small reserve stocks at this level are the major component of food security in Africa. Such household reserve stocks are of small size, are progressively consumed during the period between two harvests, are rarely carried over to more than one year, and the cost element to manage them is minimal. Before independence and especially in countries in which the urban population became quite large, the colonial administration established silos and warehouses to store grains in order to ensure regular and adequate supply to the urban mining and export crops plantation workers who were engaged in activities that prevented them from producing their required food. Large-scale food stocks were effectively established during the rapid urbanization of Africa following independence. Although the many private food marketers kept individual small or medium stores, most African governments through their parastatals chose to establish large-scale silos and warehouses in an attempt to control the greatest share of the marketed surplus. For government parastatals and private traders the rationale behind storing food was primarily commercial. The concept of building food security reserves was only given importance following the harvest failures that hit the Sahelian Zone of West Africa early in the 1970s. Constituted food security reserves were in most cases managed by public institutions and under the direct responsibility of the grain marketing boards. Although storage is one of the most important elements in food security, the experience of managing stocks in Africa has revealed constraints with respect to the appropriateness of the location and scale of food reserve stores and the predominance of food grains.

Inappropriateness of Location and Scale of Food Reserve Stores

The fact that governments decided to bear the responsibility of managing security reserve stocks has resulted in a situation in which the location of the stocks has come to be determined more by political exigencies rather than economic or operational considerations. Stocks are usually constituted either in the neighborhood of cities or in inaccessible areas that are nonetheless of interest to political leaders. Although it seems reasonable to establish reserve stocks near cities to cope with the emergency needs of the unemployed urban poor, the problem is that in cases of severe crop failures in rural areas, the stocks cannot transfer to the affected areas in time.

The ambitious large-sale silos and warehouses that have been established by governments have also brought about additional problems. In some cases the public parastatals lack the financial resources required to procure a corresponding volume of food and, as a result, the large silos or warehouses are often half-empty. The reverse is also true when the capacity of public parastatals is outstripped by huge surpluses during good harvests. Also, most of these large-scale storage facilities use modern buildings. Although such structures have the advantages of controlling temperature and humidity, there is a lack of skilled management and, subsequently, huge losses. As a result, stocks that are earmarked as security reserves are often totally lost.

Predominance of Food Grains

Food reserve stocks are now exclusively constituted of grains whereas other staples such as roots, tubers, meat, and dairy products are left out. Cereals are durable commodities and are more appropriate for long-term storage without special conditions. It has already been pointed out earlier in this book that in countries such as Nigeria, Cote d'Ivoire, Ghana, Gabon, Cameroon, Congo, Rwanda, and Zaire tuber crops contribute a larger share than cereal to the per capital calorie intake. Thus, in view of the importance of these food commodities and the production potential in Africa (95.8 million tons of root and tuber crops in 1985), they cannot rationally be excluded in any national or regional food security policy. Their perishability cannot prevent Africans from looking at ways of preserving them, especially because traditional processing methods are known to exist. The fundamental issue is whether Africa must design food security programs based on all food items produced in the continent or rely on cereal production supplemented by imports.

Implications of Food Trade in Africa's Food Security

In the past, when Africa was self-sufficient in food, food commodities were seen as biological necessities and their value was first seen in human use. With the rapid growth of total population and urbanization, food has become more and more of a commodity and its economic value is recognized more and more. Food trade at the domestic level and its political importance due to trade-offs between producer and consumer interests have already been highlighted; it is nonetheless noteworthy to add that, in practice, agricultural producers have in general received little protection; they normally receive few of the subsidies while paying heavy taxes on their exports. This discrimination has had consequences for both agricultural production and income distribution and has contributed to much of the rural poverty and hunger.

Because Africa is an agricultural primary-commodity exporter and food importer, the trade relationship of the continent with other regions will necessarily affect the continent's food policy since food security does not depend solely on domestic food production. Among the major elements of the international economy that affect domestic food policy are the balance of payments and the rate of exchange. Presently, the reduced demand for Africa's primary commodities and the depressed prices of exports have seriously increased the cost of providing food security to Africans. The food-importing countries are also input importers and this has made their domestic food outputs more expensive than world market prices and has tended to encourage them to import foodstuff. Because of their burden of debt, many African countries can hardly cope with investing in increasing domestic food production or paying for commercial imports. Therefore many of these countries are more and more dependent on massive food aid to offset domestic deficits. However, the problem is that food aid, which was originally seen as an emergency measure, has become an essential component of the food security schemes of these countries. The volume of the food aid requirement of Africa for 1988–1989 is estimated to be over 3.6 million tons.

The massive food aid to Africa has contributed without doubt to saving thousand of lives, preventing millions of people from starving, and has been used, in some cases, to promote labor-intensive developmental projects. It is, however, more and more subject to criticism including the fact that food aid represents surplus disposal by nature and because its delivery is often not planned, it has contributed to the curtailment of domestic production efforts. Also, as food aid is composed mainly of cereal grains with a great share of wheat, it has tended to affect consumption habits, which, in turn, have affected domestic production.

As far as the exchange rate is concerned, it has been established that overvalued rates act as an implicit tax on agriculture because farmers receive less for their crops than if the domestic currency was market-determined and they are charged more for imported inputs and consumer goods. Even though devaluation is a politically sensitive issue, it could to some extent be helpful in providing incentives to producers and protecting locally produced food against imports of cheap food.

Another element dictated from outside Africa that impinges on domestic food policies is the cut in subsidies (particularly for inputs and for consumer prices) called for by the ongoing structural adjustment programs. The increases in consumer prices do not necessarily correspond to increasing producer incomes but rather to increases in the margin of marketing intermediaries. Also, because such measures are often associated with stagnant or depreciated incomes of consumers and especially of urban workers, they often contribute to increases in the incidences of entitlement failures. The "hidden" subsidies provided to farmers in developed countries (e.g., huge costs of storage supported by the United States or the European Economic Community) have made locally produced food commodities less competitive than the ones imported from these countries. Although there is no doubt that developing countries, including African countries, cannot sustain the provision of cheap food to everyone, a gradual increase in the price of food commodities over a reasonable period of time is a better strategy for avoiding political upheavals such as the recent riots in several countries against "expensive food." In short, the following measures are recommended to help in ensuring food accessibility:

1. Governments should revise the existing rigid rules and regulations that have adversely affected the efficiency of the food marketing system for so long;
2. Governments' intervention in the food marketing system should not discourage private traders; rather governments should back their efforts more practically by providing the food marketing system with services that private operators cannot develop, including market information intel-

ligence, training through operational marketing extension systems, and an early warning system to cope with sudden emergencies.

If a framework involving all domestic operators is not developed in order to efficiently administer food policies, national food security objectives will remain nothing more than a political statement.

CONCLUSION

Food has become one of the most important items in any discussion of African development in the last two decades. The wish has been to find effective ways of ensuring that all Africans have access at all times to the minimum quantities of food necessary to lead active and healthy lives.

It is, however, important that food security is not narrowly defined as being simply synonymous with food self-sufficiency. Rather, it should be viewed not only as food availability through domestic production, storage, and trade, but also as accessibility to food through home production, purchase in the market, or food transfers. In this regard, a distinction has to be made between permanent and temporary food insecurity and it must be recognized that famine, hunger, and malnutrition are often as much a problem of poverty as of shortages of food supplies.

This chapter provides a brief assessment of the food security situation in Africa and describes some of the major challenges that confront African leadership with regard to strategies for increasing food and agricultural production, sustaining food production through effective resource management, and ensuring adequate accessibility to food through more effective marketing and trade policies. The essential precondition for overall social and economic growth in Africa is a dynamized agricultural sector brought about by a steady increase in agricultural productivity resulting from technological change. For African countries to get out of the food and debt trap they have now fallen into, domestic food production must increase so as to feed the growing population, and export production must rise to generate the necessary foreign exchange to pay off the continent's debt and meet essential import needs. Both these developments will require agricultural research to generate new technologies that will permit the improved levels of crop and livestock production needed to dynamize the agricultural sector.

The challenge is to develop an agricultural research and technology generation system in each African country and in the region at large that will involve many participants in a two-way process, beginning at one end with agricultural scientists carrying out their research at research stations and extending to include the farmers themselves. The range of activities involved in this process includes basic or fundamental investigations, research on experiment stations, adaptive experiments on farmers' fields, and experimentation by farmers themselves. These activities must also be accompanied by a strong national extension system as well as other agricultural support services.

Food and agricultural production per se will not guarantee food security for the peoples of Africa. The increases in the food production must come from a sustainable food production base. The two most important factors impinging on the sustainability of food production in Africa are environmental degradation and erratic weather conditions. This chapter outlines the challenges that lie ahead: to conserve the continent's soils at all costs and to minimize the vagaries of the weather.

The issue of food accessibility, in both its physical and economic dimensions, has been examined. The challenge of improving food security in Africa needs to include actions aimed at improving the adequacy of food supply and the efficiency of the distribution system of food, including storage, preservation, transport, marketing, and processing. In this regard, the challenges of the future will include finding appropriate ways of revising existing rules and regulations that have adversely affected the efficiency of marketing systems in Africa and of ensuring that any needed government intervention in the food marketing system does not discourage private participants.

APPENDIX

At the subregional and regional levels current food security projects include advisory services to the Inter-State Committee on Drought Control in the Sahel (CILSS) and the organization of a subregional early warning system accompanied by joint efforts in the reduction of food losses and the improvement of postharvest technologies in the countries of the Southern African Development Coordination Conference (SADCC). The government of Australia also undertook the identification of training needs in the area of food security in 14 African countries during 1983 and 1984.

A number of other subregional groupings have proposed the initiation of common projects or programs in the field of food security. These are the Economic Community of the Great Lakes Countries (CEPGL), the Economic Community of West African States (ECOWAS), the Preferential Trade Area of Eastern and Southern African Countries (PTA), and the Inter-Governmental Association Against Drought and Desertification (IGADD).

Similarly, the following political groupings, which include a certain number of African countries, have proposed or set up committees to discuss common activities related to food security: the Arab Organization for Agricultural Development (AOAD), which includes Egypt, Libya, Somalia, Sudan, and Tunisia in Africa; the Organization of the Islamic Conference (OIC), which includes Algeria, Burkina Faso, Cameroon, Chad, Comoros, Djibouti, Egypt, Gabon, Gambia, Guinea, Guinea-Bissau, Libya, Mali, Mauritania, Morocco, Niger, Senegal, Somalia, Sudan, Tunisia, and Uganda from Africa; as well as the

Coordination Bureau for the Non-Aligned movement, which includes 45 African countries. Most recently the adoption of a common food strategy has been proposed within the framework of the Central African Customs and Economic Union (UDEAC), which covers Cameroon, the Central African Republic, Chad, Congo, and Gabon.

In addition to the action programs and projects referred to above, there are also research programs and projects in progress in Africa that are variously funded and executed mainly by universities or similar bodies, generally from outside Africa, as well as by the International Food Policy Research Institute (IFPRI). Thus Michigan State University is engaged in a program of research dealing directly with the issue of food security in Africa financed by the United States Agency for International Development (USAID) and has instituted specific projects in Senegal, Rwanda, Somalia, Mali, and Zimbabwe. Similarly Germany, through its executing agency GTZ, is at present engaged in a comprehensive grain marketing and food security study for the Sahelian region.

National resources devoted to the improvement of food security by member States have been, and continue to be, on the whole minimal by comparison with external aid for this purpose. Apart from a very small number of exceptions like Tanzania, which has allocated since 1979 a total of Tsh. 30.9 million to its national food security programs, or Nigeria, which has established a unilateral trust fund for food security, most African countries have made no more than token contributions, generally in the form of personnel and logistical support, to their programs in this priority field.

External assistance to enable countries to implement their food security programs has been and is being provided by a large number of international aid agencies. Funding is both multilateral and bilateral. Prominent among international agencies active in the field of food security is the FAO, which, through its Food Security Assistance Scheme (FSAS), executes a number of food security projects and has been a catalyst for many more through the reports of its missions dispatched to over 25 countries in Africa to study their food security needs.

The Economic Commission for Africa is also active in this field at the subregional level and is currently executing five bilaterally funded projects. Four projects, two funded by Germany, and one each by Algeria, and Nigeria, through the United Nations Trust Fund for African Development, are assisting countries in the Great Lakes Area and in Central Africa and West Africa to reduce postharvest food losses, and in the case of the Great Lakes area, to harmonize their food security programs; the fifth, funded by Belgium and run in collaboration with the International Centre of Insect Physiology and Ecology (ICIPE), is diffusing some of the Centre's biological methods of insect control for the reduction of both preharvest and postharvest food losses in Eastern and Central Africa.

REFERENCES

Abalu, G. O. I., *Solving Africa's Food Problem: Food Policy* (1987), pp. 247–252.

ECA, *African Alternative to Structural Adjustment Programmes: A Framework for Transformation and Recovery* (Addis Ababa, Ethiopia, 1989).

ECA, "A comparative study and analysis of ongoing food security programmes in Africa." Report to the Thirteenth Meeting of the Conference of Ministers (Addis Ababa, Ethiopia, April 23–27, 1987).

Food and Agriculture Organization, *African Agriculture* (Rome, Italy, 1986).

Food and Agricultural Organization, *Director-General's Report on World Food Security: A Reappraisal of the Concept and Approaches* (Document CFS.83/4), 1983. Rome, FAO.

Food and Agricultural Organization, *Food and People* (Rome, Italy, 1985).

Food and Agricultural Organization, *Food Supply Situation and Crop Prospects in Sub-Saharan Africa: Special Report* (Rome, 1989).

Leibenstein, H. *Famine and Economic Development.* Edited by K. M. Cahil, *Famine*, 1982.

Malton, P. J., and D. S. Spencer. "Increasing Food Production in Sub-Saharan Africa: Environmental Problems and Inadequate Technological Solutions." *American Journal of Agricultural Economics* 66(1984):671–676.

Valdés A., and P. Konandreas. "Assessing food insecurity based on national aggregates in development countries." In A. Valdés, editor: *Food Security of Developing Countries.* Boulder, Colorado: Westview Press, 1981.

World Bank, *World Development Report*, New York, Oxford University Press, 1988.

World Resources Institute. *World Resource 1990-1991* (New York: Oxford University Press, 1990), p. 292.

Chapter 4

Dietary Patterns in Africa

Idris M. Nur

INTRODUCTION

Diets in Africa are the functions of ecological zones and vary among different groups. Some groups maintain an adequate diet, some face temporary shortages when harvests fail, and other are permanently malnourished. Many people have neither the means to produce food nor the money to buy it and are at the mercy of factors such as the weather, food prices, job availability, and so on. Such people are particularly vulnerable to food insecurity, largely because of poverty. Some African countries have been doing their best to remedy the situation.

The primary contribution of agriculture is food, adequacy of which is essential for nutritional health. Fruits and vegetables, such as mangoes, pawpaws, tomatoes, carrots, and others provide vitamin A as well as vitamins B and C. Green leafy vegetables, such as spinach, sweet potato leaves, cassava leaves, and others are rich in several nutrients—iron, folic acid, and others. Weaning foods can be made locally by mixing cereals with legumes or oilseeds; these could be promoted as being nutritionally beneficial.

EXTENT OF DIET DISTRIBUTION IN AFRICA

The types of diet consumed in Africa depend mainly on the ecological zones. Cereal-based diets are found in the Saharan and Sudanian zones of West Africa, in the equatorial Savanna of East Africa, and on the eastern and the high plateaus. Wheat, millet, and sometimes barley are the staple cereals of the nomads populations of the Saharan, sub-Saharan, and Sahelian zones. Seasonal calorie deficiencies exist among these nomads and animal proteins are used to meet energy requirements. Millet and sorghum are the staple food grains of the peoples of the West African savanna, from Senegal to the Nile Valley. These staples provide a satisfactory protein/calorie balance but are deficient in some vitamins. Seasonal food shortages exist in the zone also. Maize is predominant in the eastern and central plateaus stretching from Tanzania to Angola. The protein/calorie balance is fairly satisfactory in this diet but some vitamins are lacking. Rice is predominant in Madagascar and is highly preferred by the local population. In Senegal and the Gambia, rice consumption, which is rapidly increasing, is met from large imports as well as from local production in the irrigated areas along the courses of the major rivers. A predominantly rice diet has an unbalanced protein/calorie ratio and is also deficient in protein, calcium, and vitamins. Roots, tubers, and plantain diets are typical in the Guinean zone and especially in the dense equatorial forest. In some areas (West Africa) these diets are associated with rice and maize and in others (Central Africa) staple food consumption is almost exclusively comprised of starch roots. These diets are deficient in proteins and some vitamins and do not provide a balanced protein/calorie ratio. Pulse-based diets (those based on peas, beans, or lentils) are confined to the highlands around lake depressions, including Rwanda, Burundi, and certain parts of the Congo, Uganda, and Tanzania. They provide satisfactory protein/calorie balance but are deficient in some vitamins.

The unequal distribution of protein between African regions is more significant than calorie differences. Fish is, however, preferred as a readily available source of animal protein in some countries. Its availability has been small and limited to the source of supply due to distribution difficulties. The consumption of animal protein is generally centered in large urban centers rather than in rural areas. The relatively high incomes of people living in the large towns is the reason for the greater demand for food that is rich in animal protein. The problem of undernourishment is largely a function of poverty and its solution calls for a broad-based attack on the poverty problem. Additional factors that could adversely affect food distribution are related to illiteracy, social taboos, and food habits. The various dietary patterns identified above also illustrate the heterogeneity of sociocultural conditions. Steps could be undertaken to help the most deprived and vulnerable group to combat hunger and malnutrition. Such steps could include the attainment of higher incomes, instituting nutrition programs, breaking diet barriers, planting school and kitchen gardens, and improving marketing methods. Imports of food, especially wheat, have been steadily increasing in Africa. Wheat is the largest single item in the value of Africa's food imports—about 38 percent—and it is rising steadily. The demand for wheat has increased more than for other cereals. Africa's growing dependence on cereal imports account for about 51 percent of the value of Africa's food imports.

EFFORTS UNDERTAKEN IN THE AREA OF FOOD ADEQUACY

Many African countries are now intensifying agricultural production of their basic staples and supplementary foods. Food self-sufficiency has become a major goal in the development plans of a number of African countries, including the Congo, Mauritania, Nigeria, Ghana, Zambia, Sierra-Leone, Somalia, and Tunisia. Some member countries have already reached the stage of self-sufficiency, e.g., Zimbabwe is self-sufficient in maize and Botswana is self-sufficient in eggs and poultry. Nigeria, Equatorial Guinea, Sierra-Leone, Cameroon, and other countries have already established national committees for the Green Revolution. During the last two decades, research on food crops and livestock in Africa at the national and regional levels has gradually achieved greater importance and now occupies its rightful place. At the continental level, the following institutions were established:

IITA—the International Institute of Tropical Agriculture was established in 1967 in Ibadan, Nigeria.

ILRAD—the International Laboratory for Research on Animal Diseases was established in 1973 in Nairobi, Kenya.

ILCA—the International Livestock Center for Africa was established in 1974 in Addis Ababa, Ethiopia.

There have been a number of integrated schemes in Africa that were designed for pooling resources from member countries. Some of these are:

WARDA—the West Africa Rice Development Association, which is composed of 12 countries and aims at making West Africa self-sufficient in rice.

CILSS—the Inter-State Committee for the Control of Drought in the Sahel is composed of eight countries. It has helped in the production and distribution of improved seeds and planting materials. The objectives of CILSS for the year 2000 are to double the production of maize, millet, sorghum, and meat, obtain a fivefold increase in the output of rice, and reach a target of 500,000 tons of wheat per year.

SADCC—the Southern African Development Coordination Conference is composed of nine countries. It has outlined, among other things, a program aimed at achieving increased food self-sufficiency on a regional basis.

ECOWAS—the Economic Community of West African States is concerned with improving the coordination of investment policies.

Efforts made by the specialized agencies of the United Nations, international organizations, donors, and intergovernmental and nongovernmental organizations in the areas of food, agriculture, health, and nutrition were significant. For example, imported commercial weaning foods are very expensive and very scarce. Thus, promotion of local weaning foods has been the area of focus in several African countries. Such foods are sound, safe, economical, and can be made in the home or on a cottage-industry scale by individuals or groups of women. The following mixtures have had successful results: e.g.:

In Ghana:
90% millet, 10% cowpea (brown bean)
90% maize, 10% cowpea
4% maize, 40% cassana, 10% cowpea, 10% dried fish, 90% millet, 10% groundnuts
80% maize, 20% groundnuts
80% sorghum, 20% cowpea
In Ethiopia, "fafa" is composed of the following:
57% wheat
10% pea flour
18% defatted soyabeans
8% sugar
5% dry skimmed milk
1% iodized salt
1% minerals and vitamins

Research undertaken by the FAO in collaboration with the governments of Senegal and Sudan produced the technology needed for the purification of composite flours. The latter, with wheat flour as a base, can incorporate up to 30 percent of local flours—sorghum, millet, maize, and cassava—and can produce good bread that is tasty, marketable, and acceptable to the consumers. In addition cassava is a good plant source of energy. Some other countries, e.g., Uganda, initiated intensive research with the aim of promoting composite flour. Composite flours comprised of wheat and local flour in varying proportions (e.g., 50–50, 20–80, etc.) are used in the making of items such as biscuits, cakes, and pasta.

RECOMMENDATIONS

There are several programs and strategies that could be used for ensuring diet distribution in Africa. The supply of roots and tubers as a whole is likely to meet future demand without difficulty. Yams are nutritious and preferred to other tubers, but are more expensive because of the large manpower requirements to produce them, the brevity of the harvesting season, and their high perishability. A specific policy to rationalize and intensify their cultivation in ecologically suitable areas around large urban centers and along major communications routes is called for. Cassava is favored from the standpoint of calorie yield per land area but is poor in food value. The production of other tubers like sweet potatoes should be encouraged. Pulses and grain legumes should be strongly encouraged because of their high nutritional value. Sugar consumption in Africa is rather low and many African countries are in a position to grow sugar, although at prices that are higher than the world market. The supply of vegetable oils should meet demand even though the latter increases with urbanization. The setting up of dairy farms with stabling facilities near large cities should be promoted. To keep the supply deficit gap of beef, mutton, and goat meat from widening, commercializing intensive stock raising and improving sanitation and health conditions are two important steps. Guaranteeing the supply of fresh eggs at fairly cheap prices is important. With the expansion of towns and cities, demand for the supply of fresh and frozen fish will certainly grow. In the long term, most of the net supply deficit of the region could be met by imports of cheap frozen fish, some of which could be dried for rural consumption.

Plans and strategies to increase food production should reinforce efforts in the following areas:

1. improving productivity of small farmers;
2. increased allocation of resources to the agricultural sector;
3. development of indigenous food crops;
4. action-oriented research;
5. reduction of postharvest food losses;
6. improved management of farm services and inputs;
7. incentive policy package combining higher producer prices and more open and competitive markets;
8. national food security schemes, with emphasis on reserve stocks.

Part Two

The Practice of Agriculture in Africa

Chapter 5
Problems and Prospects

Thomas R. Odhiambo

The central challenge of scientific research in Africa is how it can help to mobilize and build the home-grown human capacities in the continent so as to make sensible personal, family, and national development decisions on a sustained basis: such a policy standpoint requires that there be a vigorous and continuing interaction between the research community, the policymakers, the implementors of policies, and the users of the research results. This interactive relationship is nowhere evident in Africa—not in any of the major sectors of the economic life of this continent, such as mining and manufacturing, and is only seen in a few areas of agriculture, such as the tree crop industries of tea, coffee, cocoa, and palm oil. We need deliberately to bring these four groups of actors together in a common feeding-trough in order to develop a common agenda and act out a common future for the economic well-being of the continent.

African agriculture, now at a difficult crossroads, not sure which path to follow in preparation for the even more problematic and severely competitive twenty-first century, must be able to do at least three things in a superlatively productive manner. First, it must be able to regain its capacity to provide the food and nutrition that the continent's people have traditionally expected up until two decades ago. Until then, somber biblical-scale hunger and famine were known in Asia, the Middle East, and even in Europe, while Africa only periodically experienced famine after long periods of drought that were followed by huge locust plagues. But even then the highly developed postharvest technology and village-level food security systems in most of Africa ensured

---

"Visions for IPM in Africa." Address first given in its substantial content at the Final Plenary Session at a Conference on Pest Management and the African Farmer. Duduville, Nairobi, Kenya, May 21-26, 1989.

that an organized domestic food aid system would come to the rescue of the affected communities.

Second, African agriculture must regain its capacity to provide a challenging and worthwhile way of life: an agrarian environment should not only be farmed, it should also be worked, perceived, known, and lived. In short, we are looking at farming as a way of life rather than merely a source of employment. It is this holistic approach to farming that may begin to attract the youthful and literate community back to the land, rather than an approach that relies on mere exhortation to stop the drift to the urban centers.

Third, African agriculture is likely to provide the simplest and quickest entry point for modern scientific methodology to transform African societies by building upon the existing traditional agricultural knowledge base. Everybody knows that science grows by building on knowledge that is already tested and accepted; this accumulation of small steps in the growth of science goes on all the time throughout the world. It is only rarely that startling and epoch-making discoveries that revolutionalize conventional scientific thinking and technological development are made. In the last century these have included the discovery of the unit of inheritance, the gene, and how it is characterized in biochemical terms, and the discovery of antibiotics, to mention only two. In similar fashion, the traditional knowledge base grows and develops through the accumulation of new experiences and the tested results of experimentation and repeated observation. It is evolutionary in character, but it is nonetheless solidly based even though it does not necessarily grow through the analytical methodology of science. The challenge for the African agricultural scientist is to be able to understand and rationalize this vast traditional knowledge base so as to have a systemic foundation on which to build a truly African-oriented science-based agricultural practice. To ignore the traditional African agricultural knowledge base is unscientific and wasteful of past intellectual energy. Thus, the farmer–scientist relationship we are seeking will be firmly based on a rationalized accumulated tradition body of knowledge.

There are few research centers anywhere in the world that deal with the indigenous knowledge base of African agriculture. One such institution is the International Center of Insect Physiology and Ecology (ICIPE), based in Nairobi, Kenya, which established three years ago a specific multidisciplinary group—the Social Science Interface Research Unit—to implement a program of pest management technology development through a detailed understanding of traditional African knowledge of plant protection practices. This was to be done before the program was scientifically rationalized by biologists and was upgraded to a modern practice. As recently stated by a conservation scientist of the International Union for the Conservation of Nature and Natural Resources (IUCN): "Indigenous people are the repository of understanding lost by urban and industrial societies. They illustrate the lifestyles that are sustainable through the judicious use of renewable resources."

PEST MANAGEMENT INNOVATIONS

What the farmer is seeking in Africa at this time—and probably for a long time to come—is the assurance that he will harvest, year after year, without interruption or failure, a reasonable yield of his crop, whether plant or animal, and that he will live on it and be able to make a reasonable profit by selling his surplus yield. He is not interested per se in crop breeding, soil chemistry, pest management, or any of the specialized areas of research so dear to the specialist. It has been said that the farmer is "the great integrator"; it is in his best interest, indeed in his very survival, to integrate all the technological knowledge and experiences he has acquired over the years from neighbors, scientists, or others to achieve the goal he has set himself and to ensure that he has a reasonable return on his labor and investment. If the reduction of losses caused by insects can make a significant impact on his overall performance, he will certainly want to know about it and test it fully.

Caution is a quality that every farmer exercises, but he is by no means the only one who does so. Even though innovation is the life blood of industrial development, innovators normally have a hard time selling their innovative ideas because new ideas are perceived as a threat to the existing technology or as an attack on the status quo. As stated so succinctly by an industrial watcher: "Big corporations are always reluctant to hand power to men with novel ideas, especially if they cost a lot of money. They feel more comfortable with people who understand business, who play by the rules. Ideas are all very well, but they involve risk."

Thus, an innovator must persist with the demonstration of his faith in the workability and profitability of his innovation. It is said that the founder of the McDonalds fast-food chain, the late Ray Kroc, had a sign behind his desk that read: "Nothing in the world can take the place of persistence. Talent will not; nothing is more common than unsuccessful men with talent. Genius will not; unrewarded genius is almost a proverb. Education will not; the world is full of educated derelicts. Persistence and determination alone are omnipotent."

The need to keep in mind that innovations brought to the attention of the farmer will only be accepted after clear and repeated demonstration that the new technology or idea is superior to the conventional practice tells us at least three things. First, it is prudent to bring the farmer into the researcher's world at the time of formulating a scientific research and technology development (R & D) agenda, so that he can take a personal stake in the goal to be achieved by the R & D activity. Second, the extension specialist needs to be a participant in both setting of the R & D agenda and in its implementation, so that he can deliberately bridge the perceived gap between the scientist's intellectual vision and the farmer's social vision. Third, the policymaker needs to make it his business to create the political space that will allow all the three partners in R & D—the farmer, the scientist, and the extension specialist—to have an identical goal because they will have adopted an identical R & D agenda.

That agenda can only be implemented if the three parties use and understand a common language. That does not necessarily mean that the African farmer must become a scientist, or that the extension specialist must become a researcher, although we should not brush this possibility aside. What is essential is that the importance of the scientific information being conveyed or the technological innovation being demonstrated is fully realized. There is a widespread mistaken idea that print-based information is the only type of information that is important and that anyone who is not conversant with the technology of its acquisition is illiterate and beyond rehabilitation. But Africans are highly literate in oral and traditional knowledge, in which most modern scientists are illiterate. Both sides need to accommodate each other so as to provide an environment for interactive communication and productive partnership.

THE FARMER'S IPM SCIENCE

We already know that in tropical Africa a key element in agricultural production is its interaction with insect life. The pressure of insect pestilence on crops and animals in tropical Africa is more intense than anywhere else in the world. Fully 20 to 30 percent of our harvests are lost in the field and another 10 to 20 percent are lost in storage, on the average, through the depredation of insects as pests in themselves or as vectors of plant and animal diseases. We therefore need to recognize the crucial importance of insect pests in tropical agriculture. Yet, to drastically reduce these losses the farmer needs pest management technologies that are effective, long term and sustainable in nature, simple in application, and cheap in cost.

Pest management technologies with these characteristics are not in general use anywhere in the tropics, where the largest pest problems exist. There are, however, two classes of component technologies that can form the foundation of an integrated pest management (IPM) system in tropical Africa: These are: (a) intercropping, a traditional technology that on a closer scientific look has revealed its efficacy in reducing pest pressure under certain cropping patterns; and (b) the development of crop or animal resistance or tolerance to specific pests or insect vectors of tropical diseases.

We know that of 31,000 accessions of the world sorghum germplasm repositing at ICRISAT in India, about 150 accessions have degrees of resistance to the sorghum shootfly, stem-borers, the sorghum midge, and headbugs. These resistant cultivars were collected from traditional varieties of domesticated sorghum, from many countries—Ethiopia, Sudan, Cameroon, Nigeria, India, and others. It is the farmer who discovered, bred, and nursed these resistant cultivars, but the scientific plant breeder is now using these resistant lines to transfer the resistance characteristics to other more desirable or designed sorghum varieties.

Plant breeding is a practice that the African farmer understands well so it is conceivable that he will achieve a rapid understanding of the goal of modern genetic engineering just as well. However, he may not necessarily grasp the techniques for gene transfer itself, whether through the use of a micro-shotgun or a microscope-directed laser, or the utilization of a disarmed Tiplasmid from *Agrobacterium tumefaciens*. Any of these techniques can be employed to transfer a desired gene from an isolated cell into a plant cell and then to regenerate the transformed cell into a whole fertile plant that is able to reproduce the new characteristics brought on by the implemented gene. The ability of the farmer to be familiar with such techniques is not important. What is important is that in employing genetic engineering, which will soon be as common as classical plant breeding techniques but much quicker and easier to control, we will be able, for example, to transfer a particular resistance-conferring gene to a selected plant against specific pests without environmental hazards. What the farmer will need to acquire is the seed. Africa must therefore enter the field of molecular biology and biotechnology with foresight and enthusiasm.

Equally, we must get involved in fundamental research into biological control with a deliberate view to develop technology that will be congruent with our knowledge of pest biology and will match the farmer's quest for efficacy, sustainability, simplicity, and cost effectiveness in IPM. A basic fact of the present status of biological control research is that we cannot predict if a natural insect enemy of a pest, when introduced into general environment of the pest, will get established and reduce the pestilence successfully. We have only a few guidelines to give us some hope; we have then to adopt a trial-and-error approach to the selection of an appropriate insect parasite. We have an excellent chance in the tropics for discovering the principles governing this phenomenon that is so essential for adopting biological control as a regular component of IPM.

THE VISION FOR IPM IN AFRICAN FARMING

In 1989, the ICIPE hosted an important international conference on Pest Management and the African Farmer in which it brought together selected farmers, pest management specialists, and extension scientists; this even yielded a profound lesson for us all. The tripartite forum of farmers, scientists, and extension specialists demonstrated powerfully that the time has come for such a forum and that it is a vital element in the development, adoption, and implementation of IPM. The single most important act of the May 1989 conference was the recommendation that an interactive Farmers' Forum for Integrated Pest Management be established and sustained over the years to promote the farmer–scientist-extension specialist partnership, to foster the practice of formulating and implementing a common and agreed-upon R & D agenda for IPM development, and to implement IPM approaches to crop and animal production as well as postharvest technology that recognize that losses caused by pests are an important factor in the overall agricultural performance of Africa.

REFERENCES

Cowen, R. C., "New-Wave Crops Head for the Farm," *Topic* 180 (1988): 7–9.

Davis, William, "They All Laughed," *Global Business* (Spring 1989): 18–21.

Greathead, D. G., "Biological Control as an Introduction Phenomenon: A Preliminary Examination of Programs Against *Homoptera*," *The Entomologist* 108 (1989): 28–37.

McNeedy, J. A., "Indigenous Knowledge in the Development Process: Policy and Practice of IUCN," *CIKARD News* 1(2) (1989): 2–3.

Rao, K. E. P. and Reddy, V. G., "Status of the World Collection of Sorghum Germplasm at ICRISAT," in *Collaboration on Genetic Resources* (Patancheru, Andhra Pradesh, India: ICRISAT, 1989), pp. 13–15.

Chapter 6
The Farmer's Viewpoint

Mark Chona

Born behind an ox plough and almost under a cow's udder, I have the background of a cattleman and of the ox plough. But I also have the background of irrigation, a caterpillar, and a combine harvester. I have gone through the painful problems of a young boy between the ages 5 and 13 herding cattle in the morning and going to school in the afternoon and still hoping to pass the exam. Thank goodness I did. I have a history of absolute success in intensive farming on a modest 61 acres where I have 15,000 broilers, 200 steers, 300 pigs, and many vegetables that I grow—in big sizes for institutions, in small sizes for individual marketeers. I have looked at huge fields of corn that were absolutely beautiful, but I have also known 1500 acres wilting after five weeks of drought. I have experienced liming in order to restore the pH in the soil. I have argued with government colleagues and I have come to understand what a civil service mentality actually is. When I was a civil servant I was offended when anybody said there was a civil service mentality. But actually there is one, just like there is a farmer's mentality.

Up to the 1950s we experienced the independence struggle and years of structural adjustment. We had experienced them before; we are not new to structural adjustment. We are not new to many of the things that are happening now. Odhiambo's chapter gives me confidence that African scientists at least recognize the huge reservoir, not only of energy, resources, and time, but of the capital that the African farmer has that is not yet managed properly and channeled toward high productive purposes. This has been neglected for so long in favor of high-flowing language and semantics that sound so beautiful but are meaningless to the nuts and bolts of everyday life for the farmer.

I was impressed to see in Sweden at a meeting celebrating 200 years of the Royal Swedish Academy of Engineering just how the northern farmer is beginning to push hard toward innovations that assist the small farmer. The big farmers will forgive me as I am now taking them for granted. They have no problem; they are very articulate and influential. They can argue with ministers and other people who are at the decision-making level. It is the small farmer who constitutes the largest percentage and whose land is not yet fully utilized. Those resources actually hold the promise for the future. I pay tribute to politicians because immediately after the end of the struggle for independence, during the 1960s and 1970s, they entered into a very difficult period of assessing and reforming institutions and making them relevant to the new values and objectives. This is not very easy to do without a management background; it is a new concept. Those in government had the highest status; in Britain they were knighted. It was not normal for someone in business to be knighted.

In the 1960s and 1970s, I attended a number of summits during which it became apparent that agricultural and economic issues were so boring to heads of state that they were left to officials. I was amazed at the little time given to economic issues, let alone to science and technology. This had a lot to do with the trust of the civil service in terms of listening to what the farmer was saying. You will see the difference when the farmer's interests receive the attention of the cabinets; this is because national interests dictate the kind of decisions that are actually made. So even in cabinets, the interests of the cabinet ministers, some of whom are very powerful, influence the thrust of the cabinet decisions, depending on the different pressure groups. All cabinets responds to pressure groups. Mrs. Thatcher has just become green, it is a sudden transformation, due to pressure. All cabinets are exactly like that and the small farmer does not have a hope in hell, but it is possible to speak to ministers, lobby them, and tell them about the poor quality of the stock feed. The point I want to make is that what we are actually experiencing now as farmers is the experience of all members of the private sector in Africa.

In Zambia, the indigenous private sector accounts probably for about 5 percent of the total private sector. I talked to Nigerians and asked them what their percentage was in reality. These people were not from the upper echelons of the sector, but those who were actually contributing to growth, not to trading. This group consists of a very small percentage. Is it really possible to influence decision-making at the cabinet level if you are so small? A group so small is small in taking information and advice and in getting research results from research baskets. All the results remain in the baskets and are never carried to the farmer. Later on, the farmer recognizes what is rotten and what is good in the basket; the good part is that our leaders have changed in the 1980s.

Economic summits are focusing on economic crises more than ever before. Combined with research in science and technology, this should improve the quality of decisions but the problem of establishing a delivery system to the farmer still remains. All results are written in English and some of them are very complicated. The common point is that our leaders at the present moment place a greater emphasis on agriculture and this will influence the changing face of agriculture in the future. I think that the attitude of the government and the bureaucracy will greatly influence the type of farming that we have. The literature regarding development probably can be measured in tons of volumes and has diagnosed the problem correctly. I don't think that we can add to what we already know. There is so much that has already been written, so much that the African farmer already knows. The real problem is interpreting it and establishing a link between the knowledge, the decision-making, the financing, the implementation, and the evaluation of the results. The linkage has been a real problem. I think I can turn to bureaucracy and say it is at fault, it is too strong. People are bound from head to foot by bureaucracy. What used to be bottlenecks are now bottle-caps—nothing moves to the farmer. A lot of very good projects are on the shelves, particularly those written by African experts. Sometimes foreign experts have a better ear and sometimes it is even better for me because I have made a proposition, gone to a foreign expert, and asked him to present it. I know it has been accepted, but if I presented it myself it would never have been accepted. That is the attitude that many bureaucrats have. So we have to know how to move, how to lower our position. The military people probably take a strategic or even a tactical strategic retreat by using someone else and you are a few steps forward.

Until the African decision-makers and bureaucrats recognize the immense value and knowledge that the African scientists and farmers have in common, we will never move forward. That I have experienced and I know that it needs to change. I am not going to talk about the incentives necessary for successful agriculture. For example, Zimbabwe's agriculture is successful partly because immediately after independence the Minister of Agriculture was himself a farmer. He did not disturb the extension services. He made sure that politics did not enter. He managed the extension service as professionally as possible. The same happened with Malawi. President Banda is a farmer. He makes sure that agriculture actually succeeds. You can see the interest. Malawi has one of the highest infant mortality rates in the region: 153 per 1000, Zambia's is 82, Zaire's is 100, and Zimbabwe's is 74—and yet there is a lot of small-scale production in Malawi. There is a big thrust for production but little knowledge about utilization of the food to achieve a balanced diet.

Food security is related to the utilization of whatever food is being produced. It is no good having mountains of food while people are experiencing a spiral of despair into death and destruction. In Zambia we used to produce one month's supply of wheat in a year because of the poor pricing and marketing structure. At the beginning of this decade the price went up. Now we are already producing almost 50 percent and we are beginning to say that wheat imports are not needed. Even the aid flow is beginning to move away from food imports and toward project financing.

I want to illustrate the need for pressure groups in decision-making. The United States' support for Israel is not just a matter of course; it is the result of a very strong pressure group. Unless we are actually able to form a very strong pressure group as farmers in joint partnership with scientists and technologists, we will not be able to influence decisions at the cabinet level in order to improve resource allocations for agriculture.

So where should the focus be? Not mechanization again; it was a disaster in the 1960s and 1970s. Not modernization, which is not necessarily development. Haven't we got skyscrapers decorating our skylines where lifts and water systems don't work? Is that development? It looks sophisticated. There are many other big projects toward which money flowed in the 1960s and 1970s. Now there are only scrap heaps of tractors and combine harvesters that have added to debt and a lot of other problems such as health education deprivation. Abandonment of good skills that are indigenous for sophistication has actually affected agriculture and agricultural practices in Africa. I have heard opposition voiced against organic manure. However, I would like to make a difference: organic manure brings insects only because it is wrongly done. Organic manure that is really mature cannot result in insects. It has to be well done. It is not a matter of putting some grass and cow dung and a bit of water—it has to mature. Maybe the rush to create an agricultural miracle out of production that is not chemically based has caused that kind of problem as a way of reducing fertilizer applications. I am absolutely convinced that I will not choose the way of organic manure. I will go to what I used before, either composite manure or, much more probably, green manure. I was born into this kind of culture. We used very successful crop rotation. When fertilizer came it made things easy. I did not have to go ploughing again as a young man, which requires walking from about 3:30 AM to 7 PM.

The youth and women suffer a lot on the farms. I used to start work when the cock crowed the second time. We had lunch while the oxen were having their lunch. I used to give the cattle their last supper before going to bed. We finished work at the same time. I prayed for rain so as not to have to go to the fields. That kind of practice shows that farming is not a lesser pastime. It is a very hard job and you require very strong incentives to get people into farming and ensure that they remain there. Unless the incentives are sufficiently attractive, it is like asking an investor to invest in a war zone—he won't. There have to be very strong incentives and that is my plea to the decision-makers. It is a very hard job, but a farm manager is always trying to cover up my disaster areas and I know it. When I visit my farms I am happy that I understand the mentality of my farm manager, that he will never tell me the truth like a civil servant does. The farmer has a real problem in making a decision. That is why a lot of people fear investing in farming. The risks are extremely high. You count on two good years, two bad years, and one average year so you end up with average production. Tell me among small farmers who can do a cash flow, to convince the bank manager, with all the tactics of covering up, the losses always on the lower side, running away from the tax man and from the bank manager. The best investment you can make is not in the bank because it will always depreciate. The bank manager does not get to it anyway. Those are some of the problems we actually have.

As a result, farming has been undervalued, underestimated, discouraged, and underachieved and it will continue to be so unless there is leadership. People thought I was crazy, they called me names, when they saw me in jeans and overalls they said that Mark Chona had gone bonkers. He is really suffering. But when I got my first Mercedes they said, look at him. I even had to sell my 7400-acre farm because it was on the road because, forgive me my fellow Africans, you know what PhD means? Pull him down! In the rural areas the communities are cynical. We are very cynical but we recognize success when it happens. In town cynicism takes another test of jealousy and eventually opposition. So if you demonstrate success and you are on the road as I was, my political instinct told me to sell. So I did and bought another two farms, one of which was on the road. But I made sure that my activities were well inside. It is extremely important because we are not yet used to success. Two reasons account for our successes—in the rural areas it is witchcraft and luck. In urban areas it is stealing. Frankly, you are occupied with the decisions of the cabinet, the fear of being successful and at the same time coping with advanced thinking, science, and technology. When can you actually have time to sit town and mark out a future that is absolutely sure and the results of which are sustainable? It is a catch-22 situation. But I still have hope that we need to motivate the farmers. I have hope that the governments will understand that you can't move forward unless there is both the push and pull. Investment in agriculture needs to be qualitative. If you move people, a few of us may be talkative like me, my job was marketing political ideas, good and bad together, in the political interest of course, only then can you persuade the farmer who is cynical. I don't trust figures. If you come and tell me that the nation has grown 2.7 percent I will tell you about queues in the shops, lack of transport facilities, and so on so 2.7 percent means nothing.

What matters is what is happening in the production areas. Actually I don't believe in stagnation. I believe you can only go forward or backward unless God makes another species of man.

I would like now to talk about accessibility. We have to be accessible to the farmers, not just to knowledge and all the various incentives. Actually, we do not have access to the farmer and that is why we are not influencing him. He is a very intelligent guy. For centuries he has survived under impossible conditions. He will survive again from hand to mouth and he will fight the war and win. You can develop all the missiles on earth but he will remain after that. I do not like the term "appropriate technology." Appropriate for what? Why? I just like technology. There are a number of things that I think are key. Education in the broadest sense of the term. It need not be direct because we are fighting a war. And when you are fighting a war it is a indirect approach sometimes. You have often to take a flanking position. Through education and skills—perhaps not directly related to agriculture—you open up the mind of the guy who is suspicious of you, in community development centers that are deemphasized and poorly funded, where it is possible to teach women and youth new skills that are eventually cash-generating. But above all they change the culture. I am talking about the most fundamental thing, the way of life, the way of thinking, the way of doing things. That is the most important thing. I did social anthropology at school. I hated it, but now I am using it because I am looking for water and have helped to build a lot of dams in my village. Now, I am looking for the river system. Two weeks ago I found it. It is a wind pump that has much more power that anything I have known and that is what I am going to install. We have to use low-cost systems, easy-to-handle integrated farming, agroforestry, ranching, pig production, poultry production, fish farming, which is related to poultry, energy, which is related to irrigation. Water development needs management. We need a lot of energy and we have plentiful supply of it. If a country like Wales can depend on wind power why not us? We need low-cost, high-yielding technology.

Finally I would note that UNDP and many other organizations have come into our lives, but the rural man knows nothing about them. He sees a lot of transport marked and donated by the Japanese and the Swedes, but it means nothing to him. Yet, what these organizations have is unbelievable in terms of resources and I do hope that the marriage between science and technology and between a politician who really is interest-based and a powerful bureaucrat will take place. He needs the support.

Chapter 7

The Zimbabwean Viewpoint

National Farmers Association of Zimbabwe

INTRODUCTION

In the last 25 years we have experienced severe droughts in Africa, mostly affecting the Sahel and Ethiopia in 1973 and the more widespread in 1983–1984 covering virtually all parts of Africa—North, West, East, and South. The last one was a threat to the lives of 150 million people.

These two events illustrate the fragility of the food security situation of most African countries, the slim margin for maneuver that they are locked into, together with the falling international commodity prices of the mid-1980s. These factors culminated in a continent-wide economic and food crisis. Is drought a fundamental cause of the rather chronic food shortages and lack of food security in Africa? Or is drought merely exacerbating the effects of a number of root causes? These are fundamental questions that have to be answered by those trying to tackle the problems of agriculture production and food security in Africa.

Africa's food production crisis has exacerbated since the early 1970s when per capita food production averaged less than 2 percent as against 3 percent for population per annum. The per capita food production index fell from 210 kg (in wheat equivalent) in 1970 to 179 kg in 1980. Food annual growth rates during the 1980–1985 and the 1986–1987 periods averaged 2.8 percent and 1.9 percent, respectively—a further deterioration. The agricultural imports of which food is a major component grew at an annual rate of 6.2 percent during 1980 to 1985, a rise of more than 3 percent per capita per year. For sub-Saharan Africa, food import increased from US$ 598 million in 1961–1963 to US$ 4095 billion in 1980–1982, about two thirds of which was for imports of cereals, mainly wheat and maize.

Since African political independence in the 1960s, its per capita food production has declined by about 20 percent. According to an FAO in-depth study on food imports and food aid, these have maintained food availability in Africa at a staggering level of 2100 calories per capita. If this trend persists up to the year 2000 low-income sub-Saharan Africa will experience unprecedentedly severe food deficits that would be difficult to meet by a combination of commercial food imports. In 1983–1984, sub-Saharan Africa's food import bill was US$ 5.2 billion, accounting for two thirds of all agricultural exports earnings. A projection for the whole continent puts the cost of food imports at US$ 28.5 billion at constant prices by the year 2010, against a total agricultural export earnings of US$ 12 billion at the most. The deficit of cereals alone will increase to 100 million tons, of which sub-Saharan Africa would account for 58 million tons. Although it is true that developed countries would produce enough to cover Africa's food needs, it is doubtful from the FAO's study whether the distribution and transport facilities of many African countries would be adequate to handle such quantities. This factor would further decrease food availability.

The question occurring to the small farmer is why there is this decline in food output—is it due to the fact that the farmer in Africa is lazy? This is an aspect that also has to be tackled by propounders of agricultural reform in Africa.

CAUSES OF THE FOOD PRODUCTION PROBLEM AND THE FOOD CRISIS AND PROPOSALS FOR CHANGE

From the farmer's viewpoint there are about five factors that have contributed to the present agricultural crisis and food security in Africa.

1. Inappropriate policies pursued by African countries with a bias against and neglect of the food and agriculture sector;
2. Inadequate investment in production, research, marketing, and storage and extension services;
3. Lack of technological change;
4. Environmental limitations;
5. Institutional weakness and lack of basic infrastructure.

Much has been documented about the above issues especially regarding the Africa food crisis so it is needless to repeat them here. Suffice it to mention that although Africa is the only continent where agriculture accounts for more than 50 percent of foreign exchange earnings, employments, and GDP, the amount of resources allocated to it by most national governments hardly reflects this. Normally it is at levels of less than 10 percent of the national budget. Industrial and urban development are favored at the expense of food and agricultural development. Within the agricultural sector itself the policies are also inappropriate, often favoring the large farmer who operates most of the time by using foreign capital. It is important further to illustrate the effect of this policy in terms of food security.

In most countries the majority of agricultural producers are small farmers and, more often than not, most agricultural policies have ignored their role in agricultural and rural development. Neglecting the small farmers in any agricultural policy formulation means that strategies that define the local circumstances will often be missed and the policies will often meet passive resistance because they will be top-down.

Often, African leaders have sometimes reluctantly acknowledged that small farmers have yet to realize their full potential, e.g., as active participants in the agricultural development policy process. In most instances, important decisions on agricultural policy are made in the corridors of power far from the farmers' fields. Individual small farmers too often lack awareness that they have a role to play in the national development efforts and are hamstrung by limited resources and limited room for the political autonomy maneuver issues necessary for designing appropriate policies. As a result small farmers in Africa, the producers, sometimes find it difficult to have their voices heard on public policy issues that directly affect agricultural production and rural welfare. Therefore it is often evident that agriculture suffers more than other sectors and within agriculture itself the majority of producers who are small farmers are not given a hearing or a platform in designing and participating in agricultural policy formulation. Also, this bias in policy within the agricultural sector resulted in the failure of the Green Revolution to meet the needs of the small farmers.

Lack of long-term investment resources, particularly for the small farmers to acquire various farm implements and machinery, and short-term financial resources to meet cash input purchases have accumulated in serious underinvestment in the sector and in the persistence of obsolete and primitive technology, all of which have contributed to the failure to produce food at the same rate as the population is expanding.

Institutional weakness and its unfavorable effects on food and agricultural production are present in four areas in most African countries: manpower development (both of farmers and technicians), research and extension, land tenure (conducive to farmer investment), and local institutions. Several factors have hampered the development of local institutions and the active participation of people in the activities that affect their lives. Projects to increase the incomes and improve the living standards of the rural people are often designed and implemented without involving the people the project is supposed to benefit, both at the planning stage and in decisions about what to do, how to do it, and who should do it. The formulation of coherent national policies that would enhance local initiative through institutions at the local level such as farmers clubs, producer cooperatives, rural associations, and village self-help groups will speed up development. Often policies that allow small farmers to express freely their ideas about development are not implemented because of the political framework.

The lack of adequate resources to invest in rural infrastructure and agricultural credit is largely a result of a failure to develop an appropriate institutional policy framework that brings about the gradual evolution of rural financial markets. Development strategies have more often intended to depend excessively on cheap capital provided by both multilateral and bilateral donors to build infrastructure. In so doing, mobilization of domestic financial resources has been neglected and therefore the establishment of rural financial institutions, which would provide financial intermediation for mobilizing rural savings and distributing credit for local development undertakings, has seldom received the attention it deserves.

The other issue that has to be properly addressed is that concerning the capacity of policymaking institutions. This deals with how the machinery of governments at all levels can respond, define, and anticipate socioeconomic goals, problems, and emergency situations, and assemble and analyze data. Most policymaking machineries in Africa have not yet sufficiently developed a capacity for long-term strategic thinking which would enable it to propose measures aimed at containing socioeconomic fall out of uforms, much has been the reliance on foreign expertise.

In order to sum up the problems confronting agricultural development in Africa it might be of interest to recall the views of Professor Adebayo Adedeji as expressed in "A Structural Approach to Africa's Food Crisis"; that any permanent and long-term solution to the food and agriculture crisis in Africa must address itself to the structural disequilibria that pervade the sector as well as the whole economy. In particular, the sharp bifurcation between traditional and modern sectors must be eliminated, the excessive dependence on external stimuli such as foreign expertise, foreign technology, and foreign trade must disappear, and the pervasive inflexibility in the flow of resources and factor inputs must increasingly mean domestic markets—dynamic and responsive—and not external markets. It is in the course of achieving such a structural change in the food and agriculture sector that the African Green Revolution will take place. Therefore what is needed in addressing the food crisis of Africa is a reorientation at the policymaking level so that any deliberations are addressed to the factors of institutional development that require reform while also allowing greater participation by the producer of food. Ideas on African reform on agriculture have to be local and specifically addressed to the needs of those involved in agricultural production, while particularly emphasizing the role of the small farmers, as has been the case in Zimbabwe since 1979.

THE ZIMBABWEAN PEASANT SUCCESS STORY

The fact that communal farmers are now key participants in the agricultural scene needs to be commended. The reason for this success is a number of interrelated factors that prove that, given the opportunity, land, and enough institutional frameworks, small farmers can be as productive as large farmers. This success is the result of several factors.

Credit and Input Supply

After independence, there was an opening up of credit to communal farmers to use in purchasing their input requirements. Although institutional problems remain, access to credit has contributed to improved production.

Marketing

The opening up of more rural depots meant that farmers had the outlet necessary to dispose their surplus. They now had the incentive to produce for the market because a ready market existed; this resulted in increased production and efficient use of resources. The communal farmers increased deliveries of maize and cotton in 1979 from a mere 10 percent to 50 percent of total marketed produce.

Institutional Development

The promotion of farmer organization in communal areas since 1980 succeeded in making the farmer aware of his rights in rural and agricultural development and thereby promoted his contribution to the agricultural marketing sector.

Government Policies

This factor deals with pricing and the development of infrastructure in rural areas, both of which improved the farmer's access to the markets and inputs. Because agricultural commodities were favorably priced, the farmer reacted by supplying the desired commodities. The government's policy created favorable conditions for small farmers, who responded by an increase in contribution.

Education

The provision of extension to farmers provides a forum for disseminating technical information that is disseminated to the end users. The concerted efforts made by the extension service to reach as many communal farmers as possible meant that more information reached this once-neglected sector; thus education had a positive impact on production.

In Zimbabwe efforts in the last few years have been targeted toward strengthening the six major functional components in agriculture, i.e., research, extension, marketing, input supply and credit, the small farmers' production, and regulatory policies. If one of these components is weak, instituting an agricultural development plan will be difficult. Agricultural developers should aim to strengthen the links between these six major components.

Chapter 8
The Women's Viewpoint

Mary Okelo

Women in Africa play an important role in agriculture. According to the statistics they produce 60-80 percent of the food grown and constitute 46 percent of the labor force. They also contribute to the production of cash crops. With the growing migration of men to the urban areas, agriculture in a number of African countries is increasingly being done by women. But in spite of their substantial contribution to food production, the growing productivity of women is increasingly constrained by a number of barriers. The constraints that women face are slightly different from those faced by men.

Women in Africa generally do not own land. They have no access to credit. They have very limited access to information, agricultural inputs, technology, and extension services, to mention just a few constraints. The African Development Bank recognizes the role that women play in development in general and in agriculture in particular. It believes that enhancing the effective participation of women in the development process is not only a matter of equity but one of economic efficiency, arising from the necessity to effectively employ Africa's human resources. African women comprise 50 percent of Africa's population. They make substantial contributions to all areas of the economy but especially to food production as is evidenced by the economic power of the West African market women.

In spite of all these substantial contributions, women continue to be excluded from the decision-making process. Women's economic contributions are not recorded or reflected in national statistics. Consequently, they are not systematically planned or budgeted for in the national planning process. Ignoring women's contributions, coupled with paying little attention to improving their condition, is a contributing factor to Africa's food insecurity. Africa cannot afford the luxury of ignoring its food producers because doing so does not make economic sense. African women must have access to productive resources in order to increase their agricultural and economic productivity.

It is in this regard that the African Development Bank is taking definite actions to support the role of women in food production. The Bank has undertaken and continues to undertake a number of measures to demonstrate its commitment to supporting women. For example, a Women in Development Unit has been established that acts as a catalyst to ensure that women's concerns are incorporated in all Bank projects and operations. The Bank supports its member countries to fully involve women not only in agriculture but in the process of development. In its policy dialogue with member countries, the Bank encourages these countries to formulate policies that enhance women's effective participation in development.

Agriculture is a priority sector of the Bank and 30 percent of the Bank's lending goes toward agriculture. For example, between 1985 and 1988 loan approvals by the Bank totalled US$ 5.9 billion, a third of which were for agriculture. The Bank also encourages its member countries to allocate more resources to agriculture and in 1988 several countries responded by increasing the allocation of resources by between 14 and 56 percent of public investment in agriculture. In 1988, Cameroon allocated over 26 percent of its public investment to agriculture, Madagascar 25 percent, and the Congo 24 percent. These are just a few examples.

The Bank continues to support research efforts and stresses the importance of the application of technology. Research activities are supplemented by agricultural extension. This entails providing credit and agricultural inputs to the farm. In all this resource allocation women are being targeted as beneficiaries, not as welfare recipients but as active contributors to economic development. For example, recently the Bank opened a line of credit for agriculture through the Commercial Bank of Uganda. Women were targeted as beneficiaries and 30 percent of the credit went to women. Increased productivity has already been recorded as a result of making credit accessible to women. Needless to say, the repayment record by women is very good.

To give yet another example of the Bank's commitment to support agriculture and, in particular, to support women, the Bank co-financed a rice project in the Gambia. In this project women farmers were allocated land in the scheme, credit and agricultural input such as improved seeds and fertilizers were made available, and extension services were provided. Projected rice production was between 1 to 4 tons per hectare but in 1984 the first harvest produced yields of 6 to 7 tons per hectare. In 1987 yields of over 10 tons per hectare were recorded. In that year a woman farmer won the prize for being the best farmer (out of men and women) in the scheme. The Bank continues to support agriculture with a view to achieving food security in Africa but at the same time pays attention to the special needs of women without marginalizing women. The Bank's policy is to incorporate women in the mainstream.

One of the general constraints faced by women deals with access to land. In additional African societies women had the recognized right to use the land they farmed. But with the advent of modern land adjudication, title generally goes to men as the heads of families and women are sometimes left with only a worker claim to the land they farm. This reduces women's economic security and increases their dependence on husbands and children. It inhibits women's access to credit and cooperatives, thus diminishing agricultural productivity. The land tenure system in Africa is a very sensitive issue and requires adjustment to strengthen not only women's standing but also the standing of the small holder. However sensitive this may be, it must be tackled.

Education is another constraint that women face in agriculture. The important role education plays in agricultural productivity is well known. Research has shown that having four years of primary education rather than none is associated with an increase of about 80 percent in annual farm output. There is also evidence that this percentage increase can be substantially enhanced if, in addition to the education factor, complementary inputs such as improved seeds, credit, and access to markets can be provided at the same time. But African women farmers do not always have the necessary education. In higher institutions of learning the percentage of women students in the faculties of agriculture is minimal. Very often, women in these institutions are enrolled in domestic science and home economics courses instead of in agriculture courses.

Technology is another constraint that restricts women's increased productivity. The importance of technology in agriculture is well known. African women work 15 or 16 hours a day using traditional tools and implements. Mechanization, especially in agriculture, has been concentrated on tasks done by men. Very often women's workloads have increased as a result of such mechanization. For example, when tractors are used to plow extended areas, women have more acreage to weed and thin. There is an urgent need for the adoption of labor- and time-saving devices and technologies that will increase the productivity of women's activities. Of particular note are technologies for water collection, storage, and food processing and preservation. Odhiambo cites statistics on the losses that are caused by poor storage conditions and this problem requires urgent attention.

The lack of agricultural credit is yet another serious constraint faced by women. To a very large extent, women are treated like untouchables by formal credit institutions, especially in terms of collateral, labor constraints, and bank practices. Formal credit institutions insist on security or collateral, usually in the form of land or property, before offering credit. However, in Africa very few women own land, so automatically women are excluded from credit.

Women face labor constraints because under some legal systems they are considered to be minors and do not have any contractual rights. In fact, in some countries that are not undertaking legal reforms, women continue to be classified in the same category as lunatics, children, and the mentally deficient. Certainly these constraints automatically bar women from getting access to credit.

Banks and financial institutions tend to be located in urban areas whereas women farmers reside in rural areas. Given their workloads, it is always costly in terms of money and time to go to banks for loans. Bank procedures are very cumbersome and bank forms are never specimens of clarity. It is necessary for bank procedures to be simplified in order for semi-literate and illiterate women (and men) to comprehend them. There is rampant discrimination against women borrowers in the Bank; before they negotiate their loans they are asked by the loan officer where their husbands are and who is going to sign for them. Many households in Africa are headed by women, and often there is no man available to sign the guarantees that these institutions insist on. Banks sometimes refuse to give loans to women on the grounds that women's requests are small and therefore costly and unprofitable in administrative terms. Given the importance of small-scale farmers who will always require credit in small amounts, banks will have to change and adjust their requirements.

Extension services are yet another hurdle that women face. Women have limited access to male extension farmers and sometimes male extension farmers prefer to talk to men. However, these men are not the ones involved in farming so some information is lost in the communication process. It is necessary to organize women into groups through which information can be channeled or to increase the number of female extension workers who can interact with women farmers.

Given the substantial contribution that women make to agriculture it is important that attention is given to their concerns. It is necessary to look into the needs and problems of women in order to tap their full potential and productivity. It is not enough to leave it to the governments—it is a challenge for all of us. It is not even enough to state in the Lagos Plan of Action that 25 percent of resources should be allotted to agriculture. We have to go beyond that and ensure that 25 percent actually goes to the people who produce the food. It is not enough to increase resources unless women farmers are ensured access to these resources. We can all play substantial roles in changing the current situation. There is a need to examine the current policies to see whether they are conducive to increased food productivity. It is also necessary to collect gender disaggregated data so that we know who is doing what and how much the resources are being used. At the moment, women are being cheated. They produce 60 to 80 percent of the food but they get less than 1 percent of the resources allotted to produce that food. Those ratios must be reversed. The education system must be examined. If they are in farming we have to ensure that they are prepared for this role. Our schools are where the problems start. At the moment women are disadvantaged because historically women's schools did not have science facilities so women are unable to get into higher institutions of learning because of their poor start. It is also time that African men started to talk about women's issues in public. Every time it is a woman or a few brave men who speak out for women. When are African men going to start? If they don't talk about it they cannot internalize it. That is their challenge.

Chapter 9
Agribusiness

George Moody-Stuart

INTRODUCTION

As a starting point, it is important to clarify what we are talking about when we refer to "agribusiness"—an ugly word but not necessarily an ugly subject. Agribusiness is essentially agriculture organized on a large scale, although the growers involved may well be small farmers rather than owners of plantations. It is profit-oriented but does not necessarily belong in the private sector—some very successful agribusinesses are parastatals, whose cash profits flow, at least in part, to the state or to the farmers. Agribusiness frequently involves some form of processing or at least packaging, and, because it is on a large scale, it also frequently involves foreign capital and foreign personnel.

In Africa, agribusiness is often related to sugar, oil palm, coffee, cocoa, tea, rubber, cotton, citrus fruits, pineapple, and, of course, livestock. Most of these crops need some processing to be of value; indeed some, like sugar cane, require a major processing operation.

Agribusiness companies, such as my own, make their money by developing, managing, or providing technical services to agricultural ventures in which they may have a majority or minority investment or no investment at all.

The aim of this chapter is to answer three main questions:

1. Does Africa need agribusiness?
   If the answer is "no," the other two questions become unnecessary. However, if the answer is "yes" or even "maybe," we should go on to ask:
2. What does an agribusiness company expect to get from an African government and what sort of risks will it mainly fear?
3. What can an African government reasonably expect to get from an agribusiness company and what pitfalls should it seek to avoid?

Finally, this chapter offers a couple of thoughts about new directions for the future.

DOES AFRICA NEED AGRIBUSINESS?

The purpose here is not to discuss the respective merits of small farmers and large plantations. Agribusiness can work very happily and successfully with small farmers. An example of such cooperation is the Mumias Sugar Company, which I helped to form in Western Kenya.

More than 20 years ago, the Kenyan government asked us to study the feasibility of setting up a sugar project in this heavily populated but very underdeveloped area. After satisfying ourselves about the suitability of the soil and the climate, a company was formed in 1971 with the Kenya government as the majority shareholder, with financial support from the Commonwealth Development Corporation and the British Government and a management contract with our company. In mid-1973 we started operating with a factory capacity of about 45,000 tons of sugar per annum and a cane supply that was half from a nucleus estate and half from about 3000 farmers.

Because the project worked well, we doubled its size and then doubled it again, but with all of additional cane coming from small farmers. Last year Mumias made 217,000 tons of sugar, with 90 percent of the cane coming from more than 32,000 farmers, each of whom has an individual supply contract with the company. About a quarter of a million people are now directly dependent on the project and it is one of the largest contributors to the Kenyan budget, through excise duty, company tax, and dividends. All those associated with Mumias—farmers, employees, government, shareholders, and managers—have benefited substantially from it.

One good agribusiness project does not mean that agribusiness is good for Africa, but this example perhaps demonstrates some of the main virtues of agribusiness. The most obvious is that a processing plant of this size could hardly have been successfully set up in any other way. It could not have been organized as a cooperative. It could, theoretically, have been directly financed by government and managed as a department of the Ministry of Agriculture or Industry, but the skills needed to organize an agro-industrial project on this scale are not commonly found within the civil service of any country. It is, of course, very much harder to start a new enterprise than it is to keep one running after it has been well set up.

This project certainly gained from the corporate managers' ability to raise finance, both for the original expenditure and for the later expansion. I would say, incidentally, that there always appears to be finance available for an agro-industrial project that is economically sound and where good management is assured. It sometimes takes time to put the finance together but, in our experience, it always comes in the end.

Last but not least, Mumias has benefited greatly from cross-fertilization with the sugar companies in other countries, which may be more easily obtained through an international agribusiness company than in any other way. Initially this came mainly through expatriates seconded by us to Mumias. More recently, the Kenyan staff of Mumias has been able to broaden its experience overseas.

I could quote many more examples of agribusiness successes and there are certainly people who could quote examples of agribusiness failures. Perhaps it is sufficient for now to say that agribusiness is potentially—but not essentially—a great force for good.

WHAT AGRIBUSINESS NEEDS FROM GOVERNMENT

So what does an agribusiness company expect from the government of a country in which it seeks to work?

First, it does not, or should not, expect large tracts of land at give-away prices. Some governments do not permit the ownership of land by foreigners and many agribusiness companies, including my own, fully accept this position. If security of tenure is necessary (and it usually is), let it be through leases of appropriate length or through ownership by a company controlled by local citizens, in which the agribusiness company may have a stake.

What the agribusiness company needs above all else from government is certainty and consistency, which are indeed needed by any new business, even outside agriculture. New projects are usually established on the basis of a master agreement between an agribusiness company and government. The agreement may cover items such as prices, import duties, infrastructure, security, and so on. An agribusiness venture, like a human being, is particularly vulnerable during the first two or three years of its life and failure to adhere closely to the commitments of the master agreement, particularly during this formative period, is likely to lead to disaster.

One famous (or perhaps infamous) case arose a few years ago when an agribusiness organization agreed to set up a large irrigated plantation and processing plant, while government was to build a dam to provide the irrigation water. The land development and factory were finished on schedule but there was no dam, therefore no crop, no production, no money to repay loans. Insolvency followed, with serious losses to all concerned. This was a bad failure by government, of course, but one cannot exonerate the agribusiness organization from its own failure to get government to perform. Governments all over the world need to be reminded of their commitments and nagged into fulfilling them.

In another country, a government, which was contractually committed to giving annual increases in the local sugar price in line with inflation, failed to do so for three consecutive years. Well-established sugar projects in that country were having a difficult time, but a new project rapidly became insolvent, which eventually cost the government a great deal of money in guarantees and refinancing charges.

A technical area in which agribusiness may be particularly dependent on government is the importation of new varieties of planting material. It is essential to minimize the risk of infection by proper quarantine, but we have encountered situations where the importation or multiplication of badly needed varieties has been made impossible by ill-informed or unnecessarily bureaucratic procedures.

However, I believe that most of my agribusiness friends would agree that their greatest anxieties are not connected with the possibility of the sort of failure that I have mentioned. Nor, I can assure you, do we lose much sleep about the risk of expropriation or seizure of assets or the results of a change of government. We generally take the view that, if we are doing a good job, any new government will need us as much as the old one did. Our biggest government-related worry in Africa is usually the availability of foreign exchange to pay for our services or, even more important, to pay for fertilizers, spare parts, and other essential inputs for the project.

We have developed and now manage a sugar estate on the Juba river in southern Somalia, where the sugar cane depends for about nine months of the year on irrigation. In Somalia one ton of diesel for the irrigation pumps produces enough cane to make 11 tons of sugar. It says much for the Somali's sense of priorities (and maybe something for our management) that we are able to make 40,000 tons of sugar a year there, but the nonavailability of forex in many less impoverished countries—particularly when import substitution rather than export is involved—poses a real threat to the viability of agribusiness.

WHAT GOVERNMENT CAN GET FROM AGRIBUSINESS

As regards the more interesting question of what governments can reasonably expect to get from agribusiness, at the top of the list I would put compliance with the laws of the host country. It should not be necessary to mention this obvious requirement but, unfortunately, it is. Sometimes more difficult to achieve, but also very important, is sensitivity to local customs and practices. In other words, agribusiness companies must strive to be good citizens. However big internationally, they must never, for one moment, be above the law.

Next I would put high standards of technical and managerial performance. The international agribusiness company is in a position to employ men with wide and deep experience. They are expensive and if you do not get good value for money, you have the wrong people.

It should now be the responsibility of agribusiness to maximize the involvement of small farmers, assuming that this is technically feasible. I have already quoted the Mumiras project in Kenya as an example of what can be done and there are, of course, many others. The fact is that some agribusiness companies have either been too slow to recognize the potential of small farmers or too lazy to develop that potential.

Nowadays, agribusiness companies should be fully conscious of environmental issues. Although it is very important to persuade farmers to practice proper soil conservation methods, for example, it is much easier for a large company to take the long view. Processing plants must also be environmentally acceptable in terms of water and air pollution. The weight of public opinion in Europe and America must now be sufficient to shame international agribusiness into adopting a responsible attitude.

I have already mentioned foreign exchange as a major problem for many agribusiness companies. The other side of that coin is that they should be very responsible and economical in their use of it, e.g., by using local materials wherever possible and, in the case of large operations, by making spare parts rather than importing them.

The training of local staff for technical and managerial jobs at all levels is a vital part of the role of the agribusiness company. Contrary to the view of some of my African friends, I do not believe that failure to train is often a deliberate plot to keep key jobs in the hands of foreigners. It may result from laziness, ineptitude, or even overwork. Sometimes local governments should share the blame. I have known cases where unreasonably rapid localization targets were set, with the result that the remaining expatriates were too hard-pressed to spend time on training.

In my company we are particularly proud of the complete localization of the Nigerian Sugar Company at Bacita about 10 years ago and delighted at its continuing success. At Mumias in Kenya we have two expatriates out of 14,000 on the payroll and technically excellent Kenyans heading every department.

What sorts of pitfalls should governments seek to avoid in dealing with agribusiness companies. One secret is that it is very much easier to make money by selling machinery and equipment than by managing an agricultural project. The result is that there are some agribusiness companies—and of course some machinery companies—that are very keen on setting up new projects but are much less interested in ensuring their success.

Some years ago the managing director of a very reputable machinery supply company asked us to look at the cane-growing potential of an area in a north African country. When we reported that the potential was extremely low and that we would not be interested in undertaking the agricultural development, he said that he would have to find somebody else who would. I said to him, "You cannot sell a sugar factory to those people." He replied "I would sell a sugar factory to the Eskimos if they wanted to buy one." Before you condemn that attitude, I think you have to remember that if you want to buy a car you do not expect the salesman to suggest that a bicycle might do the job equally well.

The moral of the story is not that machinery companies are dishonest but that if you want an agricultural development well done, you should entrust it to an agribusiness company with a contract that ensures its long-term commitment to the project. My strong advice would be to check the agribusiness company's previous track record and, if it is good, to tie them contractually to the project for several years after it becomes operational. Their remuneration should also be tied as closely as possible to the project's performance. If they do not like it, look elsewhere.

CONCLUSION

Having revealed some of the inner truths of agribusiness, here are a couple of thoughts about future directions. Both of these stem from a desire to reconcile two conflicting opinions, namely that agribusiness has a lot to offer to Africa but that it is often inappropriate for large areas of land to be in foreign ownership. In Britain, one of our companies has successfully developed a system of what are called farm partnerships. These arise when the owner of a farm realizes that he does not have the skill, the time, the money, or perhaps the inclination to make the maximum use of his land. He may then decide to contribute that land to a partnership with an agribusiness company, which contributes technical and management skills and perhaps working capital. The profits of the partnership are shared in proportion to the value of the inputs. Such an arrangement can, of course, only work on a reasonably large scale and must run for a number of years, probably not less than 10 years in most cases. It does involve a temporary loss of control by the landowner, but he retains a substantial interest throughout, which he would not do if he leased his land. Meanwhile the land is well used and improved until the owner resumes full control.

My last thought is also on the subject of ownership and management. For some years I was chief executive of a large, 96 percent government-owned project outside Africa that the government had decided should be run on strictly commercial lines. They appointed a board of eight directors, which included two senior civil servants, one from Agriculture and one from the Treasury. The other directors, including the chairman, were all from the private sector—successful businessmen, lawyers, and accountants who gave half a day every other month to help me to steer the business. The directors and I could, of course, have been fired at any time by the shareholders, who happened to be the government, but in practice we were allowed to run the business exactly as if it were part of the private sector. The success of this operation encouraged me to believe that some African governments might do well to let some of their major parastatals operate on fully commercial lines, without necessarily privatizing them. At the end of the day it is the management, not the ownership, that makes the difference between success and failure.

Chapter 10
The Needs of African Agriculture

M. Pierre-Victor Mpoyo

To find an indicator of the failure of African leadership to deal with the problems before it, it is not necessary to look further than the agricultural sector. The decline into subsistence farming is not the result of neglect or lack of interest and study; it is the result of inappropriate technologies, grandiose schemes, and misdirected application of resources imposed on a weak and largely disorganized agricultural sector. The other major factor behind this decline is that it is treated by planners, development agencies, and banks of agriculture as a problem that is distinct from the larger problem of nation-building.

Either directly or indirectly, the agricultural sector is intimately involved in virtually every major national economic decision made and its successes and failures impinge directly on the heavy debt-and-service burden under which the nations of Africa strain. For most of its history Africa was self-sufficient in food production. The colonial powers were singularly ineffectual in modernizing the agricultural sector. They interacted with African agriculture according to the needs of the metropolitan power and left a legacy of inefficiency and misdirection. To feed the British hunger for edible oils and fats the colonial administration began a massive groundnuts scheme in Tanganyika that involved great tracts of savanna land being ploughed into giant farms. It failed miserably and cost tens of millions of pounds. More importantly, it engendered the kind of big thinking that was to become endemic in Africa. A similar project in Casamance, Senegal failed, costing billions of francs. This type of massive-scale investment into agriculture became the darling of the new African leaders as they took over control of their economies.

Such a mega-solution to African agricultural projects is still being sought. A prime example is the recent US$ 49 million Nigerian rice project undertaken by the Hester Corporation of Mississippi, the National Grains Production Company, and the Government of Cross River State. The contract provided that at least 1000 hectares of rice would be cultivated in 1961, growing to over 2400 hectares the following year and to the full 4000 hectares in five years. This rice was to be cultivated by using the American system of airplane cultivation in which, after the ground is prepared, airplanes are used to seed, fertilize, and cultivate massive tracts of rice fields. Mechanized harvesters gather the crop and mill it for market. It will be no surprise that these harvests of rice have not yet filled the warehouses of Nigeria. This is a paradigm case of inappropriate technology being introduced on a scale beyond the capabilities of the local economic, social, and political authorities. It is equally important to examine this project to see exactly how it is practically linked to other, seemingly extraneous problems.

In order for this project to function there were certain prerequisites that had to be provided by the national and local governments. First, in the case of rice, were adequate water supplies. The government of Nigeria had issued tenders and awarded contracts for two rounds of borehole drilling projects to be dug in rural areas. Most of these boreholes were never actually dug and a major supply of water for this rice acreage was missing. In order to make this rice project an efficient producer of domestic rice and to use this rice as a substitute for the costly imports of rice arriving under the auspices of the "Feed the Nation" projects, the authorities had to issue licenses and hard-currency letters of credit to import seed and fertilizers for the project. An expatriate quota was to provide the manpower to supervise the project. Accommodation for 10 expatriates for a period of two months was provided by the local government for the feasibility study alone. The land acquisition and site clearance was contracted for by the governments. Heavy machinery was contracted for and spare parts were ordered. The governments advanced over 60,000 Naira toward the cost of the feasibility study. This project died a slow and expensive death, costing the governments over a million Naira in outlays and diverted resources. This was not the result of any wrongdoing on the part of the company, the government, or the transaction brokers; it resulted from the inability to mobilize the necessary resources for a project this size. The net gainers were the rice merchants of Bangkok who filled the shortfall of rice in the Nigerian market.

In addition to the outright cash wasted by the government, the opportunity costs to the Nigerian economy of the project were high. If this scheme had been more modest, had involved the rural population more, and had been better directed, Nigeria might not have been self-sufficient in rice but would have had at least some import substitution. There are scores of similar problems that have been badly handled. One need not even stray from Cross River State for examples. The Obudu cattle ranch was once a flourishing supplier of meat to a wide area of Nigeria. Today the herds are decimated. Nigeria has imported large volumes of bottled water from France and elsewhere while mineral-rich, clear springs in Ogoja remain untapped. Only recently Nigeria began to market bottled water produced domestically, water that satisfies the domestic market and frees vital foreign currency for development products. The same stories can be told of other states, other nations, other regions.

Africa is not alone in suffering problems as a result of agricultural development. In Eastern Europe the destruction of private land tenure has led to a massive bureaucratization of agriculture, a system that has led the Soviet Union and its allies to despair of ever reaching agricultural self-sufficiency. In the Soviet Union and Poland, for example, the tiny private farms account for a substantial share of domestic agricultural production while the state sector languishes in decay. This is why *perestroika* now is designed to implement cooperatives and self-help schemes in agriculture. Clearly too much government planning and control of agriculture does not work. In the nations of the EEC and North America, agriculture is also not working. Government intervention largely consists of establishing artificial price floors, booster payments, and production quotas. This has led to butter mountains, wine lakes, the destruction of thousands of pigs whose corpses are buried in gigantic pits, and the dumping and destruction of oranges and other fruits. These acts of folly are compounded by the ever-increasing search for productivity gains and ever higher yields—a process that adds a heavy burden of environmental pollution to the water supply and the air. Clearly, leaving agriculture to agribusiness alone is no solution.

The answer to the problems of the development of agriculture lies in involving the three major parties to agricultural investment—the government, agribusiness, and the small farmer. African agricultural development has left out of the planning process the two key elements that guarantee its success or failure. These are the small African farmer and the agribusiness entrepreneurs. The development of African agriculture must be a triangular arrangement: a composite of governmental programs, agribusiness innovation, and the motivated small farmer.

Most African farms are replete with broken-down equipment. Poor equipment can mean the difference between efficient production and inefficient production. It represents unusable capital in a capital-starved sector. A first practical step toward rebuilding farm efficiency is for the government to sponsor and provide local workshops where farmers can acquire spare parts and repair their broken machinery. There they can meet with other farmers and they can engage in productive dialogue with government agents on the improvement of the agricultural sector. Most frequently, small farmers lack the technical expertise to maintain their farm equipment properly. They need assistance. They need to be able to lease or borrow additional equipment at harvest time or when their equipment is undergoing repairs. They need this at an area near their farms. This is a key method of delivering practical governmental assistance where it is really needed. Moreover, the long-term consequences of establishing these workshops are also important. These workshops will be the kernels of cooperatives, the foci of mutual credit schemes, and the meeting places for farmers, traders, and the government.

One of the most difficult hurdles to surmount has been the question of the middleman. The middleman plays a unique role in the conduct of agribusiness. He is the interface between local farmers and the national markets. He buys produce and he sells seeds and fertilizers. His deployment of scarce capital in a rural area gives him a unique opportunity for being either good or evil. For the most part he has shown himself more interested in acquiring assets outside his own country. Agriculture is seen by him as a means of earning money abroad. He is not committed to development. Under the workshop scheme the middleman, if he is honest and his motives congruent with the needs of his clients, can still earn a profit. The government agent and the farmer will help keep him directed toward development. This workshop scheme is a powerful tool for harnessing the energies and aspirations of all three parties.

For any agricultural program to thrive, local and national governments must work together to provide the needed infrastructure—roads, water supply, electrical power, and access to markets. They should leave the business of agriculture to the private sector. The agribusiness sector should provide the technology, the distribution networks, and the market sensitivity so as to gain the highest return for the efforts of the farmers. It is probably not necessary to grow massive "cash crops" for markets abroad. It is necessary to grow foodstuff to feed a domestic population, to husband livestock to provide adequate meat supplies, and to secure a broad-based breeding stock to ensure future livestock cultivation. Subsistence farmers do not save; they cannot. Marginal farmers cannot save; their future is in hock to the seed and fertilizer merchants. Even successful farmers cannot save; the cash crops they produce are taken by middlemen to state-run marketing boards or smuggled to other markets where a temporarily higher price is available. Without domestic savings, particularly in the rural areas, there can be no development. The governments will have no incomes or revenue-producing enterprises to tax and no surplus profits to redirect toward infrastructural growth.

This is a circular problem. Without surplus and savings there are no funds to generate governmental revenues that will provide the infrastructure to generate further surplus and savings. State and local governments cannot constantly go with their begging bowls to the national treasury for funds. If the individual parts of the nation are poor and are not capable of generating revenue or accruing savings, how can the treasury replenish its coffers? By now most of the nation's patrimony is pledged to the International Monetary Fund or the World Bank to service the debt that has accumulated over the years. How then are African nations to extricate themselves from the burden of debt?

There is no one quick answer, but surely the start toward such an answer lies in harnessing the power of the African farmer by providing him with functioning local workshops. Cooperative schemes, credit unions, and bulk buying by rural cooperatives are surely steps in the right direction. Allowing these farmers to sell into a free domestic market is a major start, as is a system whereby farmers can obtain cheap credit in local currency for agricultural expansion. In countries like Zimbabwe this type of growth is expanding rapidly and now Zimbabwe can export foodstuffs to local markets elsewhere in Africa.

A key assistance to this process is the commitment of African nations to cooperate with each other to build links among them: road links, rail links, river links, air links. It is no accident that virtually all African wars have been civil wars. To fight a major war between nations it is necessary to have some basic logistical infrastructure. If there are no roads or bridges one cannot drive tanks or artillery across them. The colonial powers built a network of roads and rail links from the coastal capitals radiating inward to the interiors. Only very rarely were bridges built between nations. The French did not build bridges or roads to British colonies and only rarely between French colonies. Neither did the British build an infrastructure to French colonies. The Belgians and Spanish did not build an infrastructure at all. If African agriculture is to thrive, it must develop and expand an African market. This is the task of governments and leaders.

In the long run, import substitution is better than export promotion. Because the terms of trade for agricultural commodities suffer so badly against those of machinery and high-technology imports, the provision of a growing domestic self-sufficiency in foodstuffs will keep domestic prices low and international prices high as lower volumes are offered for sale in international markets. Because many African countries' products can be readily substituted by other African nations' products, selling them off the markets will help other African nations. Most importantly, import substitution permits the growth of "downstream" processing and further import substitution.

It is vital to nurture the small African farmer and to build a stable triangular structure to support indigenous production and domestic savings. Africa, moreover, can rely on these farmers to make a virtue of necessity, to avoid the ravages of untrammelled use of pesticides, phosphates, and the other environmental pollutants associated with agribusiness. Africa's waters can stay clean, its lands free from poisons. If one can add to this vision the notion of a people self-sufficient in the basic commodities of life, then the burdens of debt service and currency values become mere academic exercises. The solution lies in the hands of a free African farmer.

A key [illegible] in this process is [illegible] [illegible] [illegible] [illegible] with [illegible] [illegible] to build [illegible] [illegible] [illegible] [illegible] [illegible] [illegible] [illegible] [illegible] [illegible] [illegible] [illegible] [illegible] [illegible] [illegible] [illegible] [illegible] [illegible] [illegible]. The [illegible] [illegible] [illegible] [illegible] [illegible] [illegible] [illegible] The French did not [illegible] [illegible] [illegible] [illegible] [illegible] [illegible] the [illegible] more important [illegible].

[illegible] [illegible].

[illegible] [illegible].

Part Three

International Cooperation and Food Security

Chapter 11
The World Food Program Perspective

Ketema Yifru

Sub-Saharan Africa (SSA) has been at the center of world attention in the last two decades for several reasons including wars, drought, famine, debt crisis, population growth, and structural adjustment problems. If there is, however, an overriding element that has characterized SSA in the minds of the world over the past two decades, it is food insecurity.

Food insecurity in Africa is a serious problem that, according to a recent World Bank report, affects over 25 percent of the population of the region and up to 40 to 50 percent of the people in some of the most food-insecure countries (i.e., Chad, Ethiopia, Mozambique, Somalia, Zaire, and Zambia). Such a serious problem therefore deserves to be looked at closely, not only by Africans who are directly concerned, but by the international community as well.

The concept of food security has evolved greatly in the last few years with the FAO and the World Bank making major contributions in this evolution. Food security, as understood today, includes the following elements:

1. Increasing production in food-deficit countries;
2. Ensuring stability of supplies, including trade and the constitution of national stocks;
3. Measures aimed at increasing access of the poor to food supplies, including targeted feeding programs;
4. Employment-creating activities that enhance the purchasing power of the poor;
5. Market liberalization and the appropriate pricing policy for poor producers and poor consumers.

Although some controversy continues to exist regarding the relative worth of all these elements, the World Food Program views them essentially as working concepts that can be effectively turned into operational activities. The Program is about action, not about the definition of sacred unalterable concepts. Therefore, the important consideration for it in the field of food security rests on how international and domestic resources (including food and financial and technical assistance) can be assembled to address the basic causes of Africa's vulnerability to food shortages, hunger, and famine. Such measures have to look at both short- and long-term solutions and must be particularly sensitive to the widely known fact that food insecurity is, in many ways, deeply connected to poverty.

FOOD AID

In view of its potential for meeting people's immediate and direct nutritional requirements, for generating employment and income for the poor, as well as for realizing cash resources for investment through monetization, food aid has an important role to play in the efforts to eradicate food insecurity. Through these different functions, food aid can contribute to increases in production, help stabilize food supplies through reserves, enhance access to food by the poor through direct nutrition intervention schemes, and help in market development and restructuring by providing better stocks and counterpart funds.

Food aid supplied about 15 percent of cereal imports in the early 1970s, 25 percent in late 1970s, and 35 percent in the 1980s. It accounted for 41 percent of cereal imports of the region in 1987–1988. Recent cereal price increases will reduce further the capacity of SSA countries to import the food they need. Some projections indicate that SSA might have to double its food imports by the end of this century; however, its ability to import commercially will not increase.

Food aid has played a major role in the fight against food insecurity in Africa. The role of food aid becomes more evident during times of famine (1973–1974 and 1984–1985) than at normal periods, but the volume of food aid that flows to Africa is considerable even in nonfamine periods. The relatively high volume of food aid flows to SSA has raised questions of dependency. During the 1980s, cereal food aid flows to SSA increased from 2.2 million tons in 1981–1982 to almost 5 million tons in 1984–1985, but dropped to 3.7 million tons in 1987–1988. Its share in global cereal food aid also declined from 40 percent in 1984–1985 to 28 percent in 1987–1988. Recent periods of good weather and more liberalized food marketing policies should favor increased production and less need for food aid. But civil strife, which results in large numbers of refugees and displaced persons and disruption of production, tends to negate the beneficial effects of good weather. Africa's leadership must be made increasingly sensitive to the serious setbacks to the continent's development that result from political instability and military confrontations. Every effort needs to be made to create an open, enabling environment that will allow development efforts to progress and achieve long-term results. It is now becoming increasingly clear that the battle to overcome hunger, poverty, disease, and ignorance on the continent does not depend solely on getting the policies right. It also critically depends on creating the proper supportive political atmosphere. This is particularly true in terms of the extent to which international cooperation can contribute to durable solutions.

It is becoming evident that the partnership between African countries and the international community in pooling resources and efforts to address critical problems such as food insecurity is threatened whenever the political circumstances present obstacles. The extreme manifestation of this frustration of domestic and international efforts has been witnessed recently in these emergency situations when food and other supplies are not guaranteed safe passage to those desperately in need. The World Food Program has joined others in the international community to call for an international convention on safe passage for emergency supplies.

The United Nations Program of Action for Africa's Economic Recovery and Development emphasized the need for a new partnership between African countries and the international community in order for the hopes and aspirations of the program to be achieved. African countries were to remain firm in their commitment to reform while the international community would supplement the continent's efforts by providing badly needed external resources. At its midterm review in September last year, the General Assembly's Forty-Third session reached the consensus that even though many African governments had made brave efforts to carry out painful adjustment reforms, these had not always been adequately matched with the external resources needed for the reform programs to succeed fully. It is important for Africa's leadership to seek all possible means to ensure that the noble intentions expressed by the international community will be converted into meaningful international action through the provision of tangible resources. This is necessary in all fields of development endeavor—food security, debt relief, commodity trade reform, and emergency assistance.

FUTURE ROLE OF FOOD AID

Any careful analysis (based on even the most optimistic assumptions about growth in Africa's food production) of current population growth rates leads to the inevitable conclusion that food imports are likely to increase substantially in the foreseeable future. In view of the critical shortage of foreign exchange, the likelihood is that much of these imports will be in the form of food aid. Over and beyond the opportunities that food aid presents in addressing domestic economic problems, its provision can also help the balance-of-payments and debt situation of many African countries because the scarce foreign exchange resources will not need to be used for food imports.

Thus, it is important first to draw the attention of the international community and governments to back up the concerns expressed about the unsatisfactory food security situation facing millions of Africans with actions that lead to increased access to food, including the use of food aid. Second, it is necessary to draw the attention of governments and the international community to the challenges and opportunities for using food aid in a broad spectrum of modalities and with other resources so as to ease Africa's balance-of-payments crisis.

Chapter 12

The Food and Agricultural Organization's Approach to Formulating National Food Security Programs in Developing Countries

Leopold Wantisse Siry

As an overall approach, the FAO believes that national food security programs should embrace all three components of the broadened concept of food security, i.e., ensuring adequate availability of food supplies, ensuring stability in the flow of supplies, and securing access to food for those who need it.

However, developing countries differ in their resource endowments, institutional setups, and the nature of food security problems that confront them. Hence the formulation of a national food security program would need to take into account the specific circumstances of each country. A thorough understanding of the determinants of the existing food supply and consumption situation, the long-term and the short-term constraints in food security, and, in particular, the nature of the problems confronting the vulnerable groups in the country is essential to this exercise. Although no uniform policy framework or strategy can be described for all developing countries, there are nevertheless certain common issues that may need to be addressed. FAO's views on several of them are described below.

DOMESTIC FOOD SELF-SUFFICIENCY AND TRADE-ORIENTED SELF-RELIANCE

Many developing countries in their food and agricultural strategy stress the importance of improving domestic food production and achieving greater food self-sufficiency, particularly in staples, which are regarded as key contributors to food security. The stress on greater food self-sufficiency by many developing countries stems from important political perceptions or imperatives, as well as economic considerations. the desire for greater self-sufficiency is also reinforced by lack of confidence in international trade and the world trading system, particularly in view of the volatility of primary commodity markets and the prevalence of protectionist policies that limit the opportunities for many developing countries to expand their exports so as to earn the foreign exchange needed to finance food imports.

In practice, however, only a few countries fully meet all their food consumption needs from domestic production. Although the strategic importance of food, especially key foodstuffs, in developing countries should be recognized, it should be stressed that domestic food self-sufficiency in a literal sense is seldom feasible, nor would it guarantee food security in the sense of providing access to and stability of food supplies. Food production is normally subject to fluctuations because of changes in weather conditions, and unless surplus stocks are accumulated to meet supply shortfalls, there is always a need for imports. Thus the export sector, including usually the production of cash crops for export, plays an essential role as a source of foreign exchange. In addition, development of export crops may also be essential for overall growth and incomes in the agricultural sector. The main issue, therefore, is one of striking an optimum balance between producing food for domestic consumption and producing agricultural commodities and other goods for export. In striking this balance, a host of considerations will need to be taken into account and judiciously weighed both in terms of costs and benefits as well as possible available options.

PRICE POLICY

In most developing countries, agriculture is the mainstay of the economy and has a pivotal role to play in the economic and social development process. In order to accelerate economic development in these countries there is therefore a need to put in place appropriate policies and to give high priority to the food and agricultural sector. This includes allotting to agriculture an increased share of the development budget. In this respect, the role of price policy, amongst others, is crucial.

Price policy influences the level and commodity composition of production and thus determines the balance between domestic food self-sufficiency and production for exports. Guaranteed minimum producer prices can encourage growth in the production of both staple food crops and export commodities by reducing risks and opportunities, promoting greater use of purchased inputs and improved technology, as well as encouraging longer-term investments. However, prices by themselves are not sufficient to generate the required growth in agricultural production. There are a number of other factors that have to be in place in order that the price incentive policies can produce the desired results. First is the need to have in place machinery that ensures that farmers do in fact receive prices that are remunerative, whether obtained through direct government procurement or in the market. This will not be possible unless an appropriate support system for agriculture is in place. Research and extension will need to support the crops that are best adapted to the country's natural and human resource endowment. Inputs of the right kind must be available when required, together with the credit for their purchase. Storage capacity and the financial resources of intervention agencies must be adequate or expanded and other mechanisms must be put in place to ensure that price incentive policies do in fact work. Also, consumer goods wanted by farm families must be available in the local shops in the rural areas in order to encourage farmers to respond to the incentives of higher prices.

THE PRIVATE AND PUBLIC SECTOR

Few, if any, countries in the world give free rein to market forces to determine the course of their food security. Many instruments are used by governments to intervene in this process, such as direct interventions regarding food prices, consumption, and production and fiscal measures and social security systems that influence the pattern of income distribution and the degree of protection provided to the poor. The problem is essentially one of optimizing economic efficiency while ensuring an appropriate degree of social equity.

Institutional arrangements affecting the food marketing and distribution system have wide-ranging implications for the performance of the food and agricultural sector of any country. In recent years there has been a reversal of the earlier emphasis on the role of the public sector in many developing countries. Faced with financial constraints and often by the low cost-effectiveness of the public sector enterprises, the public sector plays a much smaller role in input supply and in the marketing and processing of agricultural commodities, either for domestic use or for export. Increasingly, government marketing agencies are being phased out or their monopoly powers curtailed so that there is a greater exposure to competition from the private sector.

However, the experience of some developing countries indicates that there is a need to keep an appropriate balance between the role of the public and private sectors. In several countries efforts to abolish private trade led to the breakdown of the marketing system and increased food insecurity. At the same time, experience in a number of countries has shown that without some government intervention, hoarding and speculative activity can create artificial food scarcities. Hence, in some countries that have made progress in tackling the food insecurity problem, such as India and Indonesia, a judicious mix of public and private sector activity has been deemed the most appropriate solution. Recent experience in Africa has also shown the need for some public intervention to manage occasional surpluses and to avoid their negative effect on prices and production, as well as to ensure distribution of food supplies to groups at risk in years of shortage. Here also a pragmatic approach is needed rather than a policy prescription based on any preconceived notions or any ideology. Above all, a pragmatic approach will need to take into consideration not only what needs to be accomplished but also the scope for alternative forms of organizational approaches, including cooperatives and producer organizations, and the capacity of public sector, in terms of administrative and financial resources, to do the job.

FOOD STOCKS

For over a decade the FAO, through its Food Security Assistance Scheme, has provided technical assistance to improve the stability of supplies in developing countries. Its activities comprise not only the establishment of modest national food reserves, but also the development and strengthening of storage infrastructure, creation of early warning and food information systems, and management training for operating such stocks. In general the FAO considers that the size of these reserves should be related to the lead time necessary to arrange for imports on food aid to meet domestic consumption requirements. Of course, the cost/benefit analysis of holding stocks is always necessary.

Moreover, although the establishment of food security reserves and the related infrastructure remain important, other measures can also contribute to supply stability. One possibility is to expand irrigation systems and introduce drought-resistant crops in order to reduce weather-induced production instability. Another is to hold foreign exchange reserves for food imports when food supply shortfalls occur.

Important considerations in arriving at the appropriate mix of commodity reserves and foreign exchange reserves would include costs of holding food or commodity reserves, availability of foreign exchange, size of fluctuations and probable import needs, variability in world prices, distance from the external sources of supply, risk aversion, and feasibility of exporting small and irregular surpluses when they arise. Further, as developing countries increase their production capacity, there is also increased likelihood that they will generate occasional surpluses, part of which can be used to create national food reserves. As with production policy, the FAO believes that appropriate stocking strategies will also differ considerably from one country to another depending on the size of the country, distance from the supply sources and other logistics problems, production variability, and so forth.

ACCESS TO FOOD BY THE POOR

Experience has also shown that increasing food production and stability of supplies by themselves do not guarantee food security for the poor. It is recognized that the acceleration of agricultural growth generally lessens rural poverty, but not always and not for all. Unless the poor have access to ownership of assets, especially land, and the access to other production resources such as credit, there is no guarantee that they could benefit from general agricultural growth. It is generally agreed that there is no real substitute for a better distribution of the ownership of production inputs, especially land, in order to reduce rural poverty. At the same time, the demographic profile in many developing countries is such that it is essential to expand employment opportunities outside the farm sector through rural work programs and the promotion of cottage and small-scale industries.

It should be recognized that long-term measures to alleviate poverty take time to yield the desired fruits. Thus, it is necessary to formulate short-term intervention schemes to assist poor families to improve their access to food. As one approach, a large number of developing countries have designed such short-term measures, mainly in the form of targeted price subsidies for consumers. National food security programs need to take into account such existing schemes and evaluate their effectiveness. In the event that such schemes are ineffective, there is a need to design alternative approaches that take into account the specific circumstances of each country and the costs involved.

FOOD AID

The coexistence of food surpluses in some parts of the world such as in most of the 1980s, and of widespread and chronic undernutrition and even famine in some others is now generally regarded as morally unacceptable. It is therefore often suggested that surpluses in developed countries should be harnessed to help in a major way in the fight against undernutrition. There is, no doubt, room for further improvement in the levels of food aid: for standby purposes in coping with emergencies, for meeting the needs of refugees, for filling the highly dangerous structural food deficits that are likely to persist in Africa and elsewhere, and for supporting development projects and programs, including intervention programs, to increase the access to food by the poor. However, it is important to ensure that food aid is properly integrated with the development plans of the recipient countries and with other forms of aid in order to avoid the possible risks of permanent dependency on other countries, of increasing disincentives to local production, and of altering the consumption patterns in favor of foods that cannot be locally produced.

STRUCTURAL ADJUSTMENT

A number of developing countries are undertaking adjustment and stabilization programs. Within this context the impact of the macroeconomic policies on the agricultural sector has to be fully recognized. Experience has shown that macroeconomic policies have often introduced disincentives to agriculture, sometimes inadvertently so. Examples of macroeconomic policies that have often damaged agriculture include higher protection given to the industrial sector as compared with the agricultural sector, as well as overvalued currencies that act as a disincentive to the production of agricultural export, crops, and import-competing foodstuffs. There is an overriding need to eliminate disincentives from macroeconomic policies and to remove inconsistencies between them and the agricultural development policies. Within agriculture, adjustment policies have included larger involvement of the private sector in food and agricultural marketing, greater emphasis on agricultural export crops, reduction of sectoral price distortions, and reduction or elimination of food subsidies. Overall, the thrust of the new policies toward providing stronger price incentives for producers should be welcomed. However, insufficient attention has been paid to date to the specific characteristics of the agricultural sectors of developing countries, which tends to constrain the achievement of the expected beneficial results of structural adjustment programs. There is a need for continued increased investment in agricultural infrastructures and institutional development, including in the private marketing and banking systems, in order to realize growth benefits.

The FAO is developing a new program of assistance to selected countries in the formulation of a comprehensive national food security program that will apply these perspectives as appropriate. The envisaged food security programs cannot be developed in isolation, they will need to be dovetailed with the broader thrusts of national economic and development plans, objectives, and priorities, making use of, for instance, major sector reviews and nutrition and rural development strategies where they exist. This assistance is expected to identify gaps requiring additional external support as well as to modify the existing policies and programs so as to elaborate a coherent, internally consistent approach to the development of the food and agriculture sector. Such an approach will achieve the ultimate goal of food security, which is that all people at all times should have both physical and economic access to the basic food they need.

Chapter 13
Rice in West Africa: The Role of WARDA

Eugene R. Terry

Many of the hungry people in the world live in the countries of sub-Saharan Africa. In 30 of the 35 countries in this region, population increases have far outstripped food production. In five countries, food production has actually declined.

The agrarian crisis in Africa cannot be explained by a few simple statistics. Insight into the crisis is complicated not only by the lack of accurate data, but also by a lack of knowledge about the dynamics of the subsistence systems of food production in the region. Any serious attempt to understand this complex food crisis must involve a critical analysis of human and natural resources, production factors, labor, capital, availability of appropriate production technology for the region's staple foods, and policy issues. We should also focus on the range of constraints arising from domestic and international macroeconomic developments and decisions imposed upon the agricultural sector in sub-Saharan Africa.

RICE IN WEST AFRICA

Recent studies (Delgado and Reardon, 1987; WARDA, 1988) show that rice is becoming more common in West African diets. In relation to other food grains, rice consumption is rising at a rate of 4 percent a year. This increase is second only to wheat as consumers switch away from the region's traditional staples of maize, millet, and sorghum. In the last 20 years, the per capita consumption of rice has doubled from 12 to 24 kilograms. If this trend continues, by the year 2000 rice will account for 24 percent of the cereal grains consumed in the eight West African countries known as the Sahel: Burkina Faso, Cape Verde Islands, Chad, Gambia, Mali, Mauritania, Niger, and Senegal. This prospect raises questions about the adequacy of current allocations to rice research.

At present, West Africa is meeting the escalating demand for rice by annually importing nearly 2 million metric tons into the region (see Fig. 1). Imports act in two ways to further widen the gap between domestic supply and demand. First, Africa's smallholder farmers cannot compete against the subsidized price of imported rice. Second, cheap imports fuel demand: the new consumers of rice are not just the region's more prosperous town dwellers but also include lower-income groups.

THE CHALLENGES FOR WARDA

Against this background of a food crisis in general and the precarious state of rice production systems in West Africa in particular, the West Africa Rice Development Association (WARDA) has been called on to evolve a strategy for rice research. WARDA is 1 of 13 international agricultural research centers

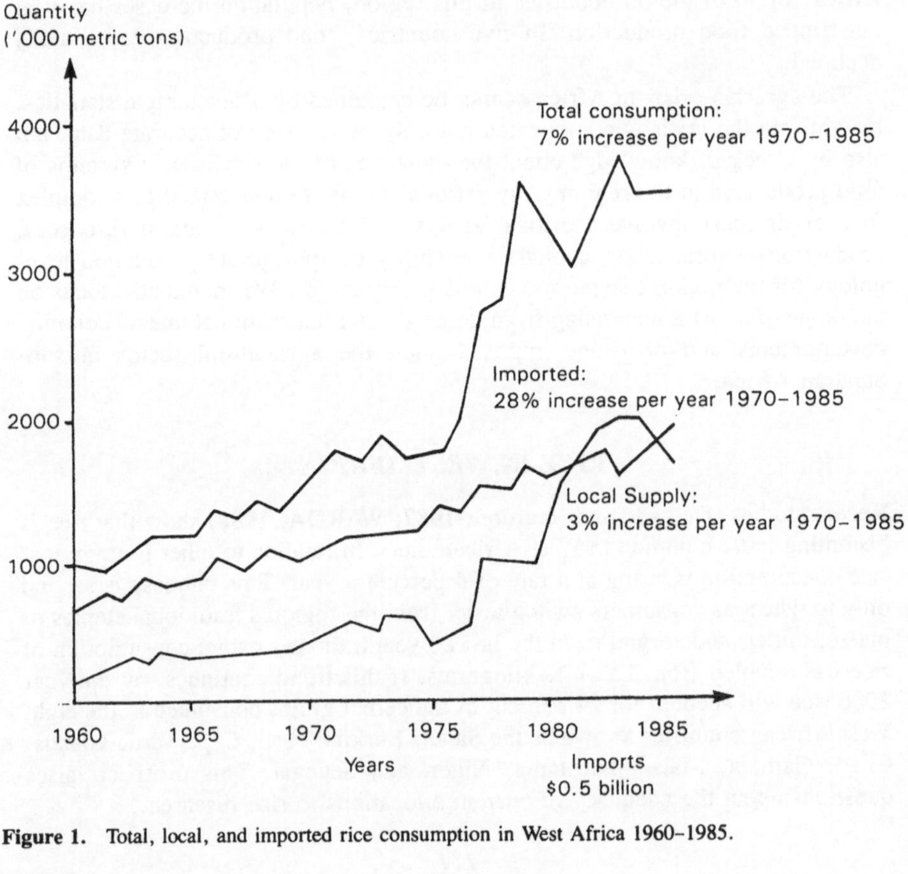

Figure 1. Total, local, and imported rice consumption in West Africa 1960–1985.

supported by the Consultative Group for International Agricultural Research (CGIAR). The challenge is a formidable one for both WARDA and CGIAR.

Evolving a strategy for rice research in West Africa poses some complex problems. Rice is grown in a wide variety of ecosystems, many of them peculiar to the region. The amount grown in any one ecosystem depends on government policies, which can change abruptly. Many of the physical and biological constraints to increased production affect all rice-growing environments; other constraints are specific to a given environment. Economic and social constraints are common across environments and affect the agricultural sector as a whole.

CONSTRAINTS

West African farmers lack appropriate tools to clear and prepare land and are unfamiliar with appropriate techniques. In the coastal zone, the shortage of cattle that is relatively resistant to trypanosomiasis (trypano-tolerant) hampers the introduction of animal traction.

Seed for rice and other cereal crops is usually broadcast. Although farmers can plant in rows using the traditional hoe, there is an acute shortage of labor to perform this task. More efficient forms of appropriate technology are needed.

Shortage of farm labor and low labor productivity limit output in nearly all subsistence production systems. Harvesting is hampered by lack of efficient tools, but mechanization is unlikely to be widely introduced in the near future. The most appropriate technology is still unavailable for threshing, drying, milling, storing, and other operations.

Macroeconomic and institutional factors affect the availability and supply of purchased inputs. One major problem is the high cost of high-technology imports. In the 1980s, imports of agricultural machinery and pesticides fell in real terms and fertilizer imports stagnated. These downward trends are likely to be accentuated by the falling prices of some export crops and the deteriorating terms of trade in general.

The most important physical and biological constraints to rice production in West Africa are those associated with the stability and sustainability of production: moisture stress, diseases, pests, and adverse soil conditions. These constraints manifest themselves through poor water retention, low and declining fertility, blast disease, infestations of stem-borers, and various conditions of nutrient toxicity. Weeds are also a major problem, often associated with inadequate management of water.

Given the limited opportunities for increasing yields by applying modern technology, rice farmers are attempting to increase production by extending the area under cultivation. In the forest zone, expansion leads to shorter fallow periods and declining soil fertility. The deforestation of upland watersheds also causes erosion, drought, and flooding, which further reduce the productive potential for rice.

Under current technical conditions, rice is probably less sustainable than other crops, given West Africa's fragile ecosystems. A major effort is therefore needed to make rice a more sustainable crop or at least to shift production to the region's more robust ecosystems. Until new technology is available to bring about a sustainable increase, governments will be forced to depend to some extent on imports. But dependence endangers national food security, as illustrated recently by sharp fluctuations in the price and availability of rice on the international market. Instability is bound to have political as well as economic repercussions.

WARDA'S STRATEGY

The national agricultural research systems in West Africa have limited facilities and personnel to deal with the many problems and constraints confronting rice production. The work of the national systems, however, is supported by WARDA, the International Institute of Tropical Agriculture in Nigeria, the International Rice Research Institute in the Philippines, and other international agricultural research centers. In addition, specialized institutes outside the region, for example, CIRAD and ORSTOM of France, are actively collecting and enhancing genetic resources and engaging in other aspects of rice research. Several institutes are concerned with increasing rice production in West Africa, but WARDA is unique because it is an intergovernmental association as well as an international center for rice research.

WARDA's task is to help national scientists and policymakers work together to offer rice farmers better technical and economic choices. Rice must take its rightful place within West Africa's mixed cropping systems, but only on certain conditions: first, the new rice technology must increase land and labor productivity and, second, it must help preserve the agricultural resource base.

WARDA's goal is to promote and strengthen the capability of West Africa's rice scientists in the science, technology, and socioeconomics of rice production. Thus put on a sound footing, rice production will improve the livelihood of smallholder farm families, increase the opportunities for rural employment, and enhance food security in the region.

WARDA recently developed its operational mandate, which ranks the types of activities that WARDA will implement to achieve its objectives. That mandate is summarized in the following mission statement: WARDA will conduct and promote research to improve the technical and economic options available to smallholder farm families, who constitute WARDA's principal target group.

WARDA's Program

Recognizing the environmental diversity of rice-farming systems in the region and the need to focus limited resources on areas where the potential impact is greatest, WARDA will concentrate on three rice-growing environments in the following order: rainfed rice systems on the upland and inland swamp continuum in the Savanna and forest-Savanna transition zones, irrigated rice systems in the belt of land along the southern edge of the Sahara desert, and mangrove swamp rice systems along coastal estuaries.

To fulfill this mandate, the research program has the following objectives:

- Develop management techniques to reduce unit production costs on a sustainable basis
- Develop higher yielding and more stable rice cultivars
- Develop means of reducing postharvest losses to economically acceptable levels
- Assess and increase the acceptability and impact of new technology
- Investigate issues affecting technology adoption
- Analyze national rice policy options

WARDA will create mechanisms to closely integrate its on-station and on-farm research. Integration will occur at all stages of technology development to maximize the value of feedback from the farm to the station. Lessons that are learned only at the final stage of technology transfer and assessment are often too late.

WARDA is also integrating research, training, and communications. Many of the national agricultural research systems in the region lack personnel who are well trained in rice research. Only 3 of WARDA's 16 member countries have more than two full-time rice scientists, another 11 have only one or two, and the remaining 2 member states have no rice scientists at all. National agricultural research systems urgently need rice scientists who are familiar with existing rice technology and know the most recent techniques for developing new technology. By integrating key components of its rice research and training activities, WARDA plans to build its collaborative research program and to contribute significantly to developing human resources in the national agricultural research system.

MEASURING PROGRESS

The extent to which research results will lead to widespread gains depends on factors beyond the influence of research institutes. WARDA's impact on regional rice production will not, therefore, be the prime measure of its progress and eventual success. On-farm trials will, however, give important clues to the potential impact of new technology.

The products of research are new knowledge and new technology; the product of communications and training activities is an enhanced national research capability. Short of having methods for quantitatively measuring these products or their effects, an important indicator of WARDA's performance will be the perceptions of its clients—the national agricultural research systems. They will let WARDA know the extent to which it has helped them acquire their own expertise to develop new technology and adapt it to the needs of farmers.

REFERENCES

Delgado, C. L. and Reardon, T. A., *Policy issues raised by changing food patterns in the Sahel* (Washington, D.C.: International Food Policy Research Institute, 1987).

West Africa Rice Development Association, *WARDA's Strategic Plan 1990-2000* (Boake, Côte d'Ivore, 1988).

Part Four
Future Directions

Chapter 14

Beyond the Crisis in African Agriculture: Balancing Conservation and Development*

Edward S. Ayensu

In the midst of Africa's agricultural crisis, it is difficult to look to the future, yet recent events have provided a better frame of reference, a chance at a more confident analysis, and a better opportunity to address the crisis successfully. I am confident that we can emerge into better times than most living Africans have ever experienced.

The new and hopeful changes in the African situation will become clear as we analyze the agricultural crisis. In a real sense, starvation is endemic, a constant, in many African countries; its speed and visibility are the only variables in this formula for human misery. A rapid onslaught makes the world take notice and recognize a famine, whereas if events move more slowly, the result is called malnutrition and remains more or less invisible outside the continent. But both kinds of hunger sap the strength of African nations.

This unhappy situation is paradoxical. At the very moment that millions of Africans were caught in the recent famine, the rest of the world was experiencing a food glut. Even when thousands of tons of grain reached African ports and airfields, delivering it to starving people presented a major, internal logistical problem to donors and recipient countries. Also in a real sense, people in some African countries are being denied a portion of the unprecedented world food supply (and the new global ability to produce food surpluses) because of a process deeply rooted in the decision-making apparatus of their own governments as well as on external influences that have always partly controlled Afri-

*Reprinted from CERES, The FAO Review 117 (20:3), May-June, 1987.

can economies. Apart from these man-made factors, there are a series of natural constraints that have influenced African agriculture and continue to do so.

The impact of human activity and of natural factors on the African environment have led to numerous assessment studies during the past two decades. Intergovernmental organizations led by the Food and Agricultural Organization, conservation groups, and private developers have participated in such studies because each group has a practical interest in the ecological balance that involves human and animal populations, the natural resource base, and the carrying capacity of the continent. In all these studies one clear note has been struck again and again: rapid population growth, coupled with the careless handling of the biosphere, have contributed significantly to the rising levels of environmental degradation present in Africa today.

It is far too easy to blame all of Africa's ills on the growing population factor. The fact remains that unequal access to good farmlands, unsatisfactory livelihood security among the rural and urban poor, and unconscionable greediness of commercial exploiters—both domestic and foreign—have driven growing numbers of people into marginal lands and have forced them to open up new forest lands. Such movement brings deforestation, soil degradation, pollution of meager usable water resources, limitation of energy supply, and reduction of food and nutritional security for many African families. And, as has been demonstrated in some parts of Africa, human conflict has resulted.

THE NEED FOR BALANCE

Those who so often refer to the enormous size of Africa and talk of their being plenty of room for expansion of human habitation delude themselves. The African continent does not have limitless carrying capacity. Forty-four percent of the land surface is arid, semiarid, and desert, 18 percent is of extremely low fertility, and 22 percent is waterlogged and of shallow depth. Only 16 percent can be considered satisfactory. Faced with this fact and the excessive rate of population growth in many African countries, government planners (especially those who continue to follow inappropriate policies) should increasingly pay attention to achieving a balance between conservation and development.

Paradoxically, the widespread use of wonder drugs such as antibiotics, vaccines for mass immunization of humans and animals, and the application of a variety of insecticides have brought considerable gains in life expectancy in Africa. It seems clear that if present trends continue, the poor populations will swell to unmanageable limits and place even more severe stress on the natural resource base. The only way out of this dilemma is to ensure that African governments themselves develop the kinds of policies that will raise the living standards of the majority of the rural and urban populations. Unless dramatic measures are taken now, the mercurial tendencies that often govern their agricultural policies and therefore their food and crop production systems will continue to jeopardize food and nutritional security.

SPECIAL ATTENTION NEEDED

Africa has a substantial endowment of resources. The continent has large petroleum and gas resources, and also commands the world's largest underdeveloped hydroelectric power potential. Every year an estimated 4200 billion cubic meters of fresh water flow out of African rivers into the oceans. The means required to catch some of this otherwise wasted water are not available. Even simple waterharvesting technologies have not been instituted, let alone the kinds of appropriate irrigation schemes that could help to offset the erratic rains experienced by African farmers year after year. The Food and Agricultural Organization has reported that by the year 2000 only 2 percent of the arable land in Africa will be irrigated. Asia, by contrast, has an estimated 34 percent under irrigation.

Africa's precious soil resources require special attention. Although our knowledge of African soils is generally poor, we do know that due to the rapid population growth in rural areas, the general practice of allowing the soil to rest and thus develop vegetative cover is becoming a thing of the past. In many parts of Africa, soil degradation progresses at an alarming rate because of indiscriminate use of the land and water resources. Furthermore, in areas in which human and animal population growth is increasing, the land is often already farmed beyond its carrying capacity—this is ecologically detrimental. In several areas some degradation is further accelerated by wind erosion, water erosion, and salinization. The rate of soil erosion has directly affected the speed with which irrigation and water development schemes become silted.

Recent climate changes in Africa have also contributed to the erosion of the natural resource base and have injected an atmosphere of uncertainty into the development process. Although no detailed quantitative studies have been undertaken to determine the extent to which human actions can influence climate, many scientists believe that the conversion of tropical forests to cropland dramatically alters the hydrological cycle. It takes only a few major timber concessions within a year to lay several thousand hectares waste by clear felling. After such commercial interests have been fulfilled, rural farmers use bush fires to complete the job. Recent studies indicate that such local conditions in the tropics affect local weather and climate even more than they do in temperate regions. Studies in the Amazon region have shown conclusively that when rain falls on an undisturbed tropical rainforest, about one fourth runs off into the ocean while three fourths reenters the immediate atmosphere to recharge the rain clouds. When the rainforest is removed by clear cutting or conversion into cropland, the ratio reverses, with three fourths of the rainfall running off into the ocean while only one fourth evaporates. There is little doubt that if present deforestation processes continue, Africa will also experience a definite alteration of the hydrological cycle—that is, if the process is not already in motion.

ECONOMIC FRAMEWORK

Apart from these environmental factors, the crisis in African agriculture must be seen within an economic framework. During the colonial period the economies of the African countries were directly incorporated into the world economy by overemphasis on the exploitation of their comparative advantage in the production of certain cash crops, such as cocoa, coffee, cotton, and timber. The African colonies in turn obtained manufactured goods to satisfy the then-small capital centers as part of a set of complex economic systems. These systems worked fairly well, so long as the colonial powers were orchestrating the whole program within the economic and political realities of the times.

With independence, most African countries began to run their economies according to a new set of rules and practices, although not necessarily with the intention of replacing the economic systems of the colonial era. In a quarter century of postcolonial history, few of these experiments have proved successful. To extricate themselves from the current system and to develop a new relationship with industrialized nations, African nations will have to make some fundamental political and economic changes within their borders to complement desired changes of policy by the external forces that indirectly influence the economic well-being of Africa's agricultural sector. Unless there are practical and pragmatic policy changes within the African countries themselves, there is no foreseeable chance that we will welcome an improvement in agricultural productivity. As an example, agricultural prices in many African countries have been kept unnaturally low because the decision-makers are more often concerned with the interests of the urban population, and tend to adopt policies favoring the cities.

In recent months (and especially during the May 1986 UN General Assembly Special Session on the Critical Economic Situation in Africa), it has been demonstrated that African governments sincerely want to establish sound foundations for sustainable and self-reliant development over the long term, regardless of the embarrassment such policy reversals might cause. Some African countries have, in desperation, subjected themselves to the harsh stabilization measures prescribed by the International Monetary Fund. These conditions have been accepted in the hope of curbing inflation through control of the money supply and achieving realistic currency devaluation, thus altering critical price-and-wage relationships. Yet even with the best of intentions on the part of the International Monetary Fund and the World Bank there are inherent difficulties in this new approach. In some countries, large-scale devaluation has caused internal price upheavals, reproducing, in effect, the same relationship between prices and wages and between internal and external prices as those that were operative before devaluation. However, if the present injection of fiscal responsibility into government policies continues and the work ethic is elevated to higher levels, a real African economic revolution could emerge. But this will happen only if the developed nations, including former colonial powers, are willing to participate in Africa's transformation for the better.

For example, the continuous influx of food aid in its current form from the developed countries is not likely to speed Africa's entry into a new era. Apart from periods of disaster, the continuous supply of food aid to countries not really suffering from famine seriously erodes their productivity. These unwise and ultimately unnecessary imports weaken the confidence of the people and undermine the economy of the continent as a whole.

POLITICAL LEGACY

Africa's economic situation cannot be fully understood without a comprehension of the political legacies involved. Most African countries have enjoyed and generally continue to enjoy their own brand of ethnic socialism, which included a fair degree of equality in the distribution of income and property. This, of course, involves the powers of tribal chiefs and, in more recent times, the preponderant power of central governments. In reality, however, Africa's political legacy dates from the fifteenth century, when European countries such as Portugal, Denmark, and the Netherlands established trading centers there. By the eighteenth and nineteenth centuries, other European countries—notably Britain, France, Germany, Italy, Spain, and Belgium—had gained possession of several territories in Africa. With the Treaty of Berlin in 1885, these European powers partitioned Africa more or less along the lines of its present political boundaries. The end result was the long history of European colonialism with its attendant actions and ramifications, including an extraordinary firm grip on exports and imports, political ideologies, and psychological influences, and the development of bureaucratic trappings devoid of meaning to most Africans.

With political independence, many African governments began to introduce their own strain of socialist rhetoric from Europe and the Soviet Union, often typified by monopolistic state-owned enterprises ridden with bureaucratic inefficiency and corruption. Opportunistic politicians (civilian and military) amassed wealth with impunity at the expense of the poor and indeed of their nations. Ironically these are the same people who use their political positions to formulate laws that prevent progress in three areas: (a) the creation of income-earning opportunities for the business community; (b) the restructuring of domestic prices of agricultural commodities and industrial goods to reflect that of international prices (or at a minimum to avoid extreme divergences of prices); and (c) the prevention of their nation from coming to grips with its limited financial resources. The economic and political structure of most African countries ignores the fact that food and agricultural production takes place in the rural areas while policies are oriented in favor of the urban constituency. As a result, the bulk of the foreign exchange earnings from the cash crops produced by the rural folk are used in the urban sectors and especially in the nonproductive sectors of the economy.

Aided by external forces, African policymakers used their political muscle to take advantage of the favorable market prices for cash crops and other marketable commodities, often at the expense of the food crops that were needed for food security. This is most neglectful of economic realities. The continuous production of cash crops has, in fact, contributed to the creation of an international glut of such commodities as cocoa, coffee, tea, sugar cane, cotton, and others. The sad outcome is that many African countries, if not most, have been placed in a situation whereby they have to import costly fertilizer, pesticides, herbicides, farm equipment, and fuel only to produce more crops that earn them less money. In effect, the production of excessive export crops has become a subsidy for the benefit of the affluent populations of the world and an outrageous surcharge upon the labors and livelihood of the impoverished and malnourished people of rural Africa. And most of this burden is borne by women.

Simply put, the attainment of political independence in the 1950s and 1960s did not improve the economic systems set in motion during the colonial era. If anything, most African countries are in worse shape today than they were at independence.

COMMUNAL TENURE DOMINATES

African agriculture has always been deeply rooted in culture and tradition. The role women play in food and crop production, for example, has always been taken for granted. In fact it is not surprising for African men to acknowledge the fact that well over 70 percent of the agricultural labor force in Africa is made up of women. Subsistence farming involving such crops as cassava, yams, cocoyam, sweet potato, plantain, and maize grown in a wide variety of agroecological zones had always involved a large proportion of women. Such farming is for many rural families a direct source of food and a source of family income with which to purchase other foodstuff and commodities. Male pastoralists have been contributing their fair share to the provision of food and purchasing power for their families.

Most African farmers depend on rain-fed agriculture. The art of establishing irrigation schemes in Asian farms have in large measure been an atypical technology for the African farmer. In recent years, however, subsistence farmers are developing this technology to ensure water availability even in areas where erratic rainfall often results in disappointing production.

Land tenure systems, with their associated customs and traditions, have always played a dominant role in African agriculture. Most subsistence farmers operate within the system of communal tenure where individual farmers use portions of land available to them. The land continues to belong to the community. However, in many instances such community lands end up being the property of landowners who exercise some form of feudal tenure whereby the subsistence farmers are given the rights to use portions of land so long as some dues are paid either in the form of labor, food, or some other reform of recompense. It is obvious that in such situations there is little protection of the rights of the subsistence farmers and for the most part they live under the shadow of uncertainty.

There is no question of the need for a comprehensive land reform program in most African countries, but taking large productive farms and dividing them into small holdings would be the wrong approach. Any transformation of existing practices should be handled with extreme sensitivity and equity because of the wide range of land tenure systems, land use policies, and the ecological conditions in most African countries.

What steps should be taken to enhance agricultural productivity in Africa as a whole? There are and there will be no shortcuts in the development process of Africa. In fact, most governments delude themselves into thinking that by some divine providence African countries will progress without the necessary steps being taken to ensure economic development.

For a start, African governments and their decision-making bodies should recognize that science and technology can make a significant contribution to the development process if they linked with appropriate government policies. This means that institutional infrastructures that further science and technology should be considered a significant component of development and as part of the productive sector of national economy. For most African countries, the highest priority should be assigned to strengthening agricultural research capability so that such institutions can help to improve and upgrade traditional farming systems. Although there has been substantial improvement in the educational programs in Africa, the fact remains that even those who receive an education are usually poorly prepared to meet the scientific and technological requirements needed to solve national and regional problems. However, if African governments and their friends abroad concentrate on the establishment and improvement of agricultural research institutions, it is most likely that within two decades every African country will have adequate numbers of trained agriculturists to ensure food and nutritional security for their countries. The Indian experience in this regard is most instructive. Some 20 years ago the government of India took a very bold political decision concerning the futility of continued reliance on food aid. With the assistance of organizations such as USAID, India established the first 12 agricultural universities, which, together with other institutions, provide that subcontinent with thousands of highly trained agriculturists. Today, regardless of other problems the country faces, India enjoys food and nutritional security.

The establishment of first-class agricultural institutions will enable the African scientific community to take advantage of the advances of the past 20 years. Marvelous things have happened in areas of cell and tissue culture, plant genetic engineering, nutrition and animal health, integrated pest management, and many other fields that can help to boost agricultural productivity, which is the only avenue through which African countries can reduce and, in some cases, eliminate poverty. This will not be an easy task. More than 70 percent of all Africans live in rural areas, despite the post-independence migration of rural populations to shanty towns in and around the major cities. Moreover, recent famines resulting from drought have launched new waves of human migration into more fortunate African countries. Under these circumstances, there is no chance to ensure constant food supplies for the growing population of Africa unless African governments make political decisions in favor of improving the productive capacity of the peasant farmers.

Because agriculture plays such a dominant role in the economic development of most African countries, it seems obvious that the pragmatic action that the governments must take should be directed toward reducing the vulnerability of the rural farmers. Simply put, African governments must invest an appropriate percentage of their foreign exchange earnings to ensure that farmers receive all the inputs they need to get their production up to speed.

African farmers have been poorly treated with respect to the pricing policies of their governments. With all good intentions, at the time of independence many African governments decided to stabilize the prices of commodities produced by the farmers to ensure that their earnings would not be affected adversely by fluctuations of world market prices. In reality the farmers have not really benefited from such stabilization schemes. Instead, the substantial foreign exchange earnings that African governments gained during the best years were essentially used to build up urban centers. In many cases large sums of foreign exchange were used to finance those ''white elephants'' sporadically distributed in the African capitals. Other substantial sums were spirited away by unscrupulous politicians.

POLITICAL IMPERATIVES

After nearly two decades of political independence, most African countries have one thing in common: they have all experienced political uncertainty. These uncertainties and outright disruptions of national life can be attributed to many factors, but first among them is the lack of political direction. Continuous changes of administrations (often accompanied by mass exists of qualified personnel from the decision-making process) have undermined the ability of many governments to establish realistic agendas and to follow them for reasonable periods. There is no question that only sustained political stability is likely to ensure proper economic planning and thus attract the business community—both domestic and foreign. Political stability is imperative if investors are to gain sufficient confidence in the system to commit their money and manpower in those countries.

Since independence the general tendency of many African governments has been to get involved in economic activities that are clearly outside the competence of governments in general. I submit that any government that is directly involved in the manufacture and distribution of soap, sardines, milk, toilet paper, and other such items is not an efficient government. The government's major preoccupation in this regard should be the creation of an atmosphere that makes it possible for the business community to operate without excessive interference or red tape.

The greatest obstacle to higher food production and general economic growth in most of the African countries has been the creation of policies that simply stifle incentive and make it impossible for the business community to perform. The development of massive bureaucracies with top-heavy governmental structures have proved to be totally counterproductive to the development process.

Careful economic planning has not been the strong point of many African governments. In a number of the countries that depend upon either one or two cash crops or other commodities such as copper, gold, or oil, there is a tendency to leap into overly ambitious development programs, especially when there is an increase in revenues. The establishment of many ill-conceived state-owned enterprises has become the hallmark of many African countries. As soon as world prices begin to fall and countries enter into acute liquidity crises, the governments become extremely unstable.

In recent times a number of African governments have been forced to develop economic readjustment programs that include substantial reduction in the overall budgetary deficit through major cuts in government expenditure, prudent external debt management, realistic examination of the performance of state-owned enterprises, and pragmatic implementation of flexible marketing and pricing policies.

Such measures are not going to be easy to implement. For a start, many African governments should give their agriculture sectors a new impetus. At present, Africa continues to suffer from poor allocation of resources; yet the bulk of the foreign exchange earnings of many countries emanates from this sector. Inadequate feeder road systems, poor extension services, poor management of the state-owned agricultural organizations, unrealistic producer pricing policies, and total disregard for serious agricultural research and support to farmers are but a few of the major factors that have resulted in the dramatic decline in agricultural production in Africa during the past two decades.

We Africans should be honest enough to recognize that poor performance on our part since independence has been a major contributing factor to our inability to feed ourselves. Our governments should not continue to blame former colonial countries indefinitely. It should be clearly understood that the developed countries of both the West and the East have their own agendas. African leadership should recognize this and learn to develop its own agendas. We must learn the game and play it exactly as the developed countries play it. Otherwise, most of Africa will risk an even more devastating and destabilized economic and social future.

Chapter 15

What Is Going Right with African Agriculture

Paul Harrison

INTRODUCTION

There are three problems confronting African agriculture: declining production, increasing rural poverty, and environmental degradation. But it is important to recognize and to emphasize that over the last four or five years several things have started going right. The country's overall performance and the prospects of individuals have shown success in increasing production, increasing the incomes of small farmers, and conserving soil, on which all future production depends. Success is important for two reasons: (1) It is good for morale. It make people realize that failure is not inevitable. If they make enough effort of the right kind they can succeed. They do not have to depend only on faith, because they also have hard evidence. (2) If success stories are looked at in detail, the reasons for the success can be discovered and general lessons, which can be applied to other efforts, can be learned.

That is precisely what I tried to do in my book *The Greening of Africa.* While researching the book, I visited 16 of the most successful projects in the field of agriculture, forestry and environment.* I also visited the major research centers—IITA in Ibadan, ICRISAT's Sahelian Center in Niger, the International Livestock Center for Africa in Ethiopia, and the International Council for Research in Agroforestry in Kenya.

The aim of this chapter is to share some of the findings and encouraging examples of that research and of more recent visits to Rwanda and Lesotho.

---

*How was success defined? Basically, I asked major donors involved to select those projects they thought were most successful and replicable. I then examined these to make sure they were replicable across a wide range of African conditions.

A RECAP ON WHAT WENT WRONG

Basically, Africa's special problems arose because of Africa's special characteristics. Western experts and Western-trained African experts went to work in Africa on the assumption that Africa was like the West, or like Asia. But Africa is Africa and is like neither.

The African environment is different in that rainfall is more unpredictable and, outside the wettest areas, drought risk is high. In addition, pest incidence is high. A single animal disease—trypanosomiasis—shapes the agriculture of most of the continent. Soils are often problematic; many African soils are easily eroded.

The African farmer is different as well. He or she if often short of cash and will avoid any risk that can be avoided because there are so many that cannot be avoided. He or she has uncertain tenure of land, and therefore has less interest in innovations that pay off only after a long time. He or she has a shortage of labor due to lack of cattle for draft power and the frequent absences of the male heads of households in cities, plantations, or mines. Finally, because of that absence, he is very often a she.

African farming is different in that two bases of agricultural advance in Europe and Asia—mixed arable and cattle farming and irrigation—are problematic in Africa.

1. Irrigation is not present on a wide scale except in small areas, where it is rather a matter of swamp farming or gardening from wells. Nor is the potential for irrigation high.
2. The potential for mixed farming, which combines cattle and arable farming, is limited because the tsetse fly bans large livestock from vast areas.
3. African crops are different: the main crops on which most research had already been done—rice, maize, and wheat—are not widely grown, and where they are, the pests are different. The basic research in plant breeding that had already been done in Asia and Latin America cannot easily be transferred to the African context.
4. African farmers intercrop a great deal, i.e., they grow a large variety of crops in one field. There are very good reasons for that.

The African government is different. Apart from the above, one of the major reasons for intercropping was that prices were unfavorable to farmers. Factors contributing to this problem are the weakness of farmer organizations and the strength of urban masses and elites. In addition, Africa reached independence with a lower level of literary than Asia or Latin America, which created problems in the recruitment of public servants. Africa had a smaller manufacturing base and a higher dependence on commodities than other parts of the developing world. So African government budgets tended to be less stable than elsewhere. As a result, programs that depended on stable and well-trained government employees or on high or stable government spending have often failed soon after donor finance was withdrawn.

I have taken the time to spell out these special characteristics, because failure to understand them has been the main source of failure of government and donor efforts in African agriculture and the reason why much of the research has not been adopted. Conversely, success is usually the result of understanding these constraints. This means introducing technologies that: demand little investment in terms of hard cash; cost little in terms of lost production (e.g., through cropland lost to nonproductive purposes); involve little risk, or indeed reduce risk; pay off the investment over a very short period; use little labor during peak work times, although they may use a fair amount of dry season labor; are also low cost in terms of government expenditure and reliance on government personnel; and, perhaps most important of all, have attractive prices.

Let us look at some examples to see how they fit the model.

Zimbabwe Maize production is an example of the conventional approach, which resulted in increased maize production among communal farmers in Zimbabwe and was, without doubt, a spectacular success. Maize yields rose from less than 1 ton per hectare to over 10 tons per hectare among people with very small farms and no private title.

Some of the features of the approach taken are generally applicable. First, it increased prices for farmers. This has really been the key. Without price incentives to produce more, farmers will not produce more food for the market, whatever other help is given them. Improved prices have been successful in boosting food production in most of the African countries that have introduced them. African farmers have shown that they are just as adaptable, responsive, and innovative as farmers of any other continent given the right policy.

Second, this approach focused on the smallholder. It has always been tempting for Western experts to recommend ''rationalization''—concentration of land into larger farms. Zimbabwe's smallholders—without having private title to their land—have disproved this theory once and for all. They have proved that the smallholder can achieve yields that are just as high as the biggest commercial farmers.

The formation of farmer's groups has also been very important. These are voluntary groupings of 20 to 40 farmers, self-chosen to make sure that they work well together. These groups have a number of functions.

They pool their labor. Instead of every family having one person look after two or three livestock, one or two people can look after all the cattle belonging to the group. Instead of every family sending someone into town to buy pesticide or kerosene, one person goes in for the whole group. Working in groups increases morale and increases work productivity. Groups are able to undertake tasks, e.g., terracing a farm—that would demoralize individual farmers.

New technologies spread more quickly in groups because group members see more of each other and discuss their techniques together.

The groups act as pre-cooperatives. They band together to make purchases and hire transport. They are also credit groups and can offer group guarantees for loans. This is important in Africa where most farmers do not have titles to land to use as collateral.

The above elements can be applied by any African government. However, other features of Zimbabwe's success are rather specific and perhaps should not be used as a model yet.

The technical package applied is expensive. For example, the technical package may involve the application of 500 kg of fertilizer per hectare of maize. This fertilizer is produced in Zimbabwe using hydroelectric energy and has been since the world isolation of the illegal Smith regime. It also involves the distribution of hybrid seeds every year. Zimbabwe has had its own breeding program for several decades. The maize variety involved, SR 52, was bred in 1949 in what was then Rhodesia. Other factors involved are a well-staffed extension service, an efficient farm credit system because credit is needed in the early years to pay for the high-cost package, and an efficient private sector system to deliver seeds and fertilizers.

Most of these elements were already in place in Zimbabwe before independence, created by and for white farmers for their own convenience. But they are not present in most other African countries yet. Of course African countries need all these things and should continue their efforts to build them, but to rely at present on an approach that demands all these elements to be in place would be to invite failure or to delay results for several years.

There are less conventional examples of a kind that could be applied by any African country almost immediately and that would have a rapid impact on food output and soil conservation.

STONE LINES INTRODUCED IN BURKINA FASO

The problem that had to be addressed was that from 1975 onward, large areas of Central and Northern Burkina Faso had begun to rapidly turn to desert. By 1985 the problem was dramatic—many villages had lost a third of all their lands over 10 years. The land was literally dead—no grass and all shrubs and even trees dead. The reason was that the soils of the area very easily form crusts when they are exposed to rainfall. Normal vegetation prevents this process but gradual overfarming and overgrazing bared more and more land until the crusts formed and no rain could get in. Even in years of good rainfall, rain would just run off, forming gullies.

The High-Tech Solution That Did Not Work

In the early 1960s the European Development Fund began to build low dykes, less than a foot high, across the fields. These dammed back the flow of water and slowed down runoff. The dykes were introduced without any consultation with farmers. They were built by bulldozers that often trundled without warning or explanation across the landscape. Most farmers had no idea what they were doing or what the dykes were good for. The dykes did not work well: they reduced the amount of water available for crops just downhill for them for a couple of meters and were unpopular with farmers. They were not maintained and footpath and cattle tracks broke through them. In some cases farmers deliberately dug them up—sabotage is a very good indicator of the unsuitability of projects.

The Low-Tech Solution That Did Work

Oxfam began work in the area in 1979, but it took them two or three years to get onto the right lines. They first wanted to introduce water harvesting—planting trees inside semi-circular miniature dams on gentle slopes where they would collect the water. But farmers said that they could not spare the water to keep seedling alive during the dry season. They wanted something that would help them with their food crops.

Peter Wright, the Oxfam worker, worked with the farmers. Traditionally, farmers piled heavy stones in lines to mark field boundaries and to slow down water in gullies. They noticed that where the lines were across the slope, crop yields were much higher. The slopes were so gentle that they are difficult to identify without a theodelite, but doing so is too technical and too expensive. Wright developed a cheap and easy technique: a thin transparent hosepipe filled with water, attached at both ends of wooden stakes. These stakes were marked at half-centimeter intervals. Two farms hold one stake each, and stand about 10 meters apart and move the stakes around until the water level is the same for each. When this happens, the bottoms of the two stakes are level. Sticks are hammered in to mark the spot and a line is drawn in the soil between them. The stones are piled along these lines, about one foot high. They dam back the rain water and give it time to soak in. This works even on the dead and crusted areas, which can thus be reclaimed for agriculture. The holes in the line let some water through, so that the area just below the lines does not get dry. This technology has proved so popular that it has spread to most of the area affected by the problem. Farmer demand is as sure a sign of success as sabotage is of failure.

How It Was Spread

Dissemination is as important to success as invention. Oxfam spread the technique by holding training courses for a few farmers from each village. They were then given one or two hosepipe levels, and perhaps a wheelbarrow or pickax or two, and went back to their village to train other farmers. This approach has worked very well in Burkina Faso. It means using skilled personnel not to train individuals directly but to train people who then train other people. It spreads the knowledge of a few skilled personnel and multiplies their influence by a factor of about 15. The information about the technology of stone lines has been spread by two other organizations, the Naam Movement, a voluntary grouping of about 1400 village societies, and the government extension service, which included information about stone lines to its central messages in 1986. This technique was so successful because it demanded little or no cash investment. The hosepipe level costs only $6 and can be shared by a whole village. It involved little loss of land—perhaps half a meter every 20 meters, or 2.5 percent. It involved no risk: farmers are not being asked to plant their whole field with a variety that might fail. Indeed, as it turns out, the lines reduce risk because in times of drought they trap a far higher proportion of rainfall into the soil of crops to use. The payoff period for the investment in labor was very short, probably well under a year if the labor cost was assessed. The average increase in yields on fields that had the lines was 50 percent, and even more in dry years. Building the lines does require quite a lot of labor, but this can be less in the dry season when it does not conflict with farm labor needs. The organizational factors that helped it to work were varied.

It was flexible. It did not set out with a rigid plan that had to be adhered to even if the whole concept proved mistaken. Within two years it completely shifted from being an agroforestry project to a project that boasted crop production and soil conservation. It made this change in response to popular demand; in other words it responded to what farmers were saying and did something about it. Additionally, it involved a high level of farmer participation. The farmers' demands determined the direction of the project and because farmers were very closely involved in developing the technology, it suited their circumstances and they were more likely to adopt it. Farmers were the main grassroots level trainers and they did all the work involved in installing the lines.

SOIL CONSERVATION IN KENYA

When I was researching *The Greening of Africa,* almost half of all smallholder farms in Kenya had been terraced to conserve soil. Traditional techniques were improved on by researchers who worked together with local people. The main technique used is the Fanya Juu Terrace. This technique was an improved version of one already used traditionally by the Kamba people in the dry area of Machakos. It involves digging a ditch along the contour and throwing the soil uphill, where it forms a miniature dam. Soil then washes down from the back of each field section to the front, so that level bench terraces are formed. While this technique is widely used, women's groups do most of the work.

Investment cost is low, risk is reduced, and payoff is quick. Increased yields of 30 to 65 percent are reported. Although labor requirements are quite high, the work can be done in the dry season. In my view, the labor involved could be greatly reduced, e.g., by greater use of trash lines, which are piles of small branches staked in along the contours with weeding trash piled against them.

WINDBREAKS IN THE MAJJIA VALLEY, NIGER

This is an area where wind erosion is a serious problem, where wind also topples crops and increases plant transpiration and water needs. There is also a shortage of fuelwood. The windbreak project, introduced by CARE, planted 40 kilometers of windbreak. These were mainly Neem trees, in double lines 100 meters apart, perpendicular to the direction of the wind. The project is so popular that farmers in areas not yet covered by windbreaks are demanding them. In some cases they are planting their own without waiting for the project.

Here again we have: low investment costs, risk reduction, and low labor requirements—the only labor is for digging holes for the trees.

There are two interesting features about this project, which are exceptions that prove the rule. In this case formal participation was not high, but a particularly gifted local forester held long evening talks with farmers to discover their needs. They told him that their most serious worry was about the effects the wind would have on soil and crops. The project was designed to meet this, their biggest priority. The amount of land lost to the windbreaks was not small—12 to 15 of every 100 meters. But careful studies have shown that the overall output of food crops per hectare increases by 18 to 23 percent where there are windbreaks, due to reduced wind erosion, reduced toppling of crops by wind, and reduced transpiration. This increase is on the whole land, including the area under windbreaks. So the Majjia farmers are getting significantly increased food production—and the fuelwood output of the trees is a pure bonus. This project could have been improved further if the windbreak trees were multipurpose, e.g., if they produced food or fodder as well as fuel. It has to be emphasized that the reason the farmers like the project is not because of the increased food output, but because of the fuelwood.

There are a few more encouraging ventures, two of which are from a different but related field: energy conservation.

Improved Charcoal Stoves in Kenya

All previous attempts to introduce improved stoves in Africa failed to get more than a few thousand improved stoves in actual use in households.

This project sold 110,000 stoves in its first two years and created a new self-financing industry that is still selling improved stoves at no cost to government or donors. Although from a different sphere, this project follows the same ground rules: the technology is low cost—the stove cost only 60 shillings, which is the cost of the old type of stove. This is a fairly modest amount and the new stove lasts six times as long as the old. Users report that it saves up to 50 percent of fuel. At prevailing charcoal costs, it pays for itself within one month in Nairobi, where charcoal is dearer, and in two or three months elsewhere. It was designed with a very high level of farmer participation, this time in the form of testing early designs in 600 households and taking note of all users' problems. It used an interesting and very effective form of dissemination: the market. The old Jiko was being made by artisans and sold by market traders; the project used the same network of artisans and traders to sell the new stove. The market can be an effective channel for other forms of innovation, e.g., new seeds and fertilizers, and may often be more effective at that than extension services would be.

High-Yielding Fodder Grass on Terrace Edges in Rwanda and Lesotho

The colonial powers forced farmers to build terraces in both countries, and to plant the terrace edges with grass. In Lesotho government conservation workers still build wide terraces planted with a grass that is not much use. These techniques were unpopular with farmers because they lost land that could have been used for food production. Farmers often ploughed the grass strips down to thin lines or to no more than a foot or less than half a meter. In Rwanda extension workers no longer expect farmers to build terraces, but instead encourage them to plant high-yielding Napier grass along the contours. This has thick roots and holds the soil back, and gradually terraces form themselves. Meanwhile the grass itself can be cut for fodder. A similar species, Bana grass, has just begun to spread in Lesotho, and once again farmers are demanding all they can get of it. In this case no land is lost to useful production. Vetiver grass is being pushed now for this purpose. But it is no use for fodder. I would argue that in most of Africa it might prove less popular than species that could be cut for fodder.

GENERAL LESSONS THAT CAN BE DERIVED

All attempts to introduce the technologies that Africa needs to increase production and conserve soil should follow the above principles, focusing primarily on low-cost, low-risk, low-labor content approaches that provide a quick return. These approaches are not only low risk and high return for farmers, they are also low cost and high return for government.

Conservation is often one of the cheapest and quickest ways of boosting production, promising increases in yields of 20 percent to 50 percent. The reason this is possible in Africa is precisely because of Africa's uncertain rainfall: conserving soil reduces the water that runs off or evaporates, and therefore increases the amount of water available for corps to tide them over dry spells. It also reduces losses of fertilizer and fertile soil.

Conservation should always be seen as an opportunity to increase production of food, fodder, and fuel, and not just as a way of conserving soil. Conservation methods that are pure cost will not succeed, e.g., stone terraces, wide grass strips planted with grass of low productivity, and state forests that local people cannot gain any benefit from. Even if extension workers pay people to install these type of things, the farmers will not maintain them afterward. They will let the stone terraces fall down and go in and cut trees from the state forests. Conservation methods in Africa should combine conservation with increased output at low cost, e.g., methods such as Napier grass lines across the contour and terrace edges planted with useful trees providing fruit, nuts, and fodder as well as fuel.

Government services should be cut and reshaped to deliver low-cost appropriate technologies. There are two main issues here: how to identify the successful technologies and how to disseminate them rapidly.

Technology Identification

The basic principles outlined above should be used as the starting point in looking for technologies that will improve production and conservation. They should inform the practice of research stations. New crop varieties can serve all the above principles, but if they are to do so they must be selected under the right conditions. If they are to be low cost to the peasant they must be able to do at least as well as his own traditional variety: without fertilizer, without additional watering, with poor management and weeding, and when intercropped according to normal peasant practice. If they are to be low risk, they must do at least as well as traditional varieties do in years of bad rainfall or bad pest infestation.

Special links must be forged between researchers and farmers. On-farm research has usually been the method to ensure that technologies that work well on the station also work well on the farm. But shortage of research staff may make this a slow business. For this reason it may help to use the extension service to do farm research. The extension service can also to some extent fill gaps in research by reporting on the practices of the best 10 percent of farmers and by identifying those local varieties that are most popular among farmers and helping to spread them. This is what has happened in Rwanda with bananas, where no improved varieties were bred on research stations. Technology dissemination should occur.

Among farmers and through traders, spontaneous dissemination should be increasingly important. When technology is appropriate to the farmer and a market network exists that can reach the farmer, the market can and will spread the best elements of those technologies that can be transported and sold, e.g., seeds, fertilizers, pesticides, and tools. The market cannot, however, be effective with technologies that cannot be transported or sold, i.e., with techniques or ways of doing things. For this it is essential to use the extension service.

THE EXTENSION SERVICE

This should be an all-purpose service that functions at the village level covering crops, livestock, agroforestry, and conservation, with specialist backup at the district level. Extension workers are often most effective if they work through farmers' groups—voluntary groupings of 30 to 40 farmers who pool labor and tools. They can meet the extension workers on one of their fields or they can elect from among themselves a village extension worker—a sort of barefoot extension worker—who will transmit the message to his fellows.

The flow of information within the service should be two-way—not only the passing on of instructions agreed upon at the central or district office, but the listening to and reporting back of farmers' needs, priorities, and problems, including problems with new technologies.

The African Green Revolution has already begun. Perhaps it might help to view it in terms of stages. The first stage focuses on all the low-cost, low-risk approaches outlined above. This creates a production boost of 20 percent or more, generates confidence in the extension service, and gets the easiest and quickest gains first. Later, as incomes rise and government services improve, factors that cost more can be added. Despite all of Africa's special problems, it is possible to increase food production and improve conservation, often at the same time, and at a cost that even poor African governments can afford.

Chapter 16

Development, Food, and Agricultural Sufficiency: The Korean Experience

Jai-Suk Chung

INTRODUCTION

In general economic theories, agriculture is seen as playing an essential role in the early development stage. Not only does rising agricultural productivity free rural labor for urban factories, rural savings constitute a major source of investment funds for the nonagricultural sector. Rising farm labor productivity also contributes to an overall increase in farm output, thereby supplying more and cheaper food for urban consumers. It further increases the general incomes of farmers so as to create a domestic market for industrial products. According to conventional economic wisdom it is often assumed that economic development should begin with agriculture and should include an attack on rural poverty. In sum, agricultural development is a primary precondition for economic growth, i.e., it has to precede industrialization.

This model, however, does not apply to the last 40 years of Korean development. On the one hand, the contribution of agriculture to the economy has been modest at best. On the other hand, the farm populace benefited greatly from the nonfarm sectors of the economy. Rapid industrialization and urbanization increased the demand for agricultural produce while providing job opportunities outside of agriculture for rural youth. Agriculture, therefore, did not in any way lead Korean economic development. On the contrary, Korea is a living example of how industrialization could precede and induce agricultural development, rather than the reverse.

With this model of development in mind, I would like to address the topic of agricultural production and food security in Africa. By way of introduction, I will offer an overview of Korea's economic development because any discussion of Korean agriculture cannot be made without reviewing the success of the overall economy. I will then discuss the role of agriculture in the context of industrialization and examine the question of food and agricultural sufficiency in Korea. I will conclude my chapter with a few personal observations on some lessons of the Korean experience.

KOREA'S ECONOMIC DEVELOPMENT: AN OVERVIEW

Korea has been transformed from a predominantly agricultural economy to a nation on the brink of becoming an industrialized and affluent society. In 1945, when Korea was liberated from the Japanese rule of 36 years, about 17 million Koreans were living in the south, most of whom were tenant farmers toiling on tiny plots of land and living below the subsistence level. As of 1988, 42 million Koreans are enjoying average per capita incomes of $4040, with three fourths of them living in urban areas. Korea is also the tenth largest trader in the world. As Scalapino aptly observed, "Today, the Republic of Korea is a medium-sized power, casting its economic shadow over a widening arc of the world."

For the sake of convenience, Korea's economic development from 1945 can be divided into five phases.

Phase 1:	Disorganization	1945–1953
Phase 2:	Reconstruction	1954–1961
Phase 3:	Economic Takeoff	1962–1971
Phase 4:	Growth Acceleration	1972–1978
Phase 5:	Structural Adjustment	1979–present

One should not assume that the drama of Korea's economic growth began suddenly in 1945. Korea has a long historical heritage that influenced its subsequent development: the tradition of Confucianism placed a high value on education, Korea's cultural identity derived from a long history as a unified country, and the legacy of Japanese colonial rule from 1910 to 1945.

Phase 1: Disorganization, 1945–1953

The first phase of disorganization was started by Korea's sudden liberation from Japan after World War II, aggravated by the division of the country along the 38th parallel, and desperately damaged by the Korean War of 1950 to 1953. Suffering economic dislocation and social turmoil, the Korean government initially achieved little in the way of economic development, except perhaps for the sweeping land reform during 1949 to 1952. In this regard, Mason (1980) and others observed: "While land reform may temporarily have reduced productivity per acre, this was far outweighed by the political and social consequences to Korea's later modernization."

Phase 2: Reconstruction, 1954–1961

The Korean War caused over one million civilian casualties and destroyed most industrial facilities and fixed assets, estimated to be well over US$ 3 billion. In terms of annual average between 1953 and 1955, the primary industry accounted for 45 percent of the GNP while the share of manufacturing was only 9 percent, implying that the economy was predominantly agricultural. Fixed investment was only 7 percent of the GNP and 90 percent of this modest figure was financed by foreign aid; this indicated the virtual absence of domestic savings.

Given the war damage and abject poverty, the most urgent task of the government was to provide people with daily necessities and to rebuild the war-torn economy. Thanks to the generous foreign aid totalling $2.3 billion from 1953 to 1961, the economy registered an average annual growth rate of 3.9 percent. Economic policies during this period were "inward looking" in the sense that the stress was on price stability rather than growth and on import substitution rather than export. This philosophy was based on the philosophy of "low exchange, interest, and tax rates."

Phase 3: Economic Takeoff, 1962–1971

Korea entered the 1960s with one of the lowest income levels in the world, a bleak prospect for economic progress, and clouded by political and social unrest. It was against this background that the First Five-Year Development Plan was launched in 1962 with an ambitious growth target of 7.1 percent during 1962–1966. One fundamental change in basic economic policy was made in a package of surgical reforms in "the fiscal–monetary–trade nexus of policies" under the principle of cost realization, which reversed the previous inward-looking policy. The new policy was implemented in succession: (a) fiscal reform to promote government savings from 1963 onward, (b) exchange rate and trade regime reforms to boost export in 1964–1965, and (c) interest rate reform to encourage private savings in September 1965. As Kim and Roemer (1979) concluded: "One consistent basic policy goal has dominated all the succeeding five-year plans: export-oriented industrialization and growth."

The spurt of Korean economy during this period was quite phenomenal, as can be seen in Table 1. In short, Korea's economic takeoff was achieved in the 1960s.

Phase 4: Growth Acceleration, 1972–1978

Spurred by the economic takeoff during 1962–1971 and an average annual GNP growth rate of 8.8 percent, Korean economy further accelerated its growth to 10.4 percent per year as commodity exports expanded by 40 percent every year. More strikingly, staple food sufficiency and some surplus in balance of payments were attained for the first time in 1977. In 1978, domestic savings rose to

Table 1. Records of Economic Takeoff

	1953–1961	1962–1971
GNP growth rate	3.9	8.8
Per capita GNP growth rate	1.2	7.2
Growth rate of manufacturing	11.7	18.5
Growth rate of export	10.7	32.4
Investment ratio to GNP	12.3*	25.2†
Domestic savings ratio	6.0*	15.4†

Sources: Economic Planning Board and The Bank of Korea
*Covers 1961
†Covers 1971

26 percent and gross investment stood at 31 percent of the GNP. It should be noted here that the government promoted the Saemaul (New Community) Movement from 1971 in order to increase farm income and to spur rural development. From 1973, to attain industrial deepening, the government invested in the heavy and petrochemical industries.

For all the remarkable gains of the 1970s, however, the management of economic policies from 1976 to 1978 were far from optimal in terms of the balance and stability of the economy. Not only did the government fail to stem the swelling flow of dollars from the Middle East construction, it also underrated the mounting pressures from domestic demand, largely due to the excessive investment in the heavy and petrochemical industries. The need for a thoroughgoing overhaul of the economy was urgent, and the painful phase of economic restructuring followed.

Phase 5: Structural Adjustment, 1979–Present

In April 1979, the government adopted integrated stabilization policies to initiate the structural adjustment of the economy. The second oil crisis in 1979 and the political shakeup led to negative economic growth of 5.2 percent in 1980. The government steadfastly sought to secure price stability by all available means. It enacted the Fair Trade and Anti-Monopoly Law in 1981, executed the zero-based budgeting in 1982–1983, and pushed for an active import liberalization policy from 1983 onward. The result has been more than gratifying because the price trend in terms of CPI has been below 3 percent from 1983 to 1987, which is in sharp contrast to 19 percent, which is what it was during the preceding four years. The balance of payments also improved with the burgeoning surplus in the current account: $5 billion in 1986, $10 billion in 1987, and $14 billion in 1988.

Recently, the Korean economy appears to have regained the momentum of rapid growth, averaging 12 percent growth a year since 1986. Hopefully, Korea will soon join the ranks of lower-level OECD economies. In spite of domestic labor disputes and rising social demands for democratization, the economy may be at the threshold of its next phase, one of transition to maturity.

INDUSTRIALIZATION AND THE ROLE OF AGRICULTURE

The preceding overview of Korea's economic progress reveals that it can be characterized as a continuous process of industrialization through its initial import substitution and export promotion, led by the expansion of manufacturing industry. As a consequence, substantial changes have taken place in the structure of national production. For instance, the growth rate of value-added in manufacturing, which averaged 11.7 percent a year during 1953 to 1961, increased to 19 percent during 1963 to 1978, and was 11 percent even during the slow growth period between 1979 and 1988. In sum, Korean manufacturing industry grew at 15 percent per year on average between 1953 and 1988.

As a result, since the end of the Korean War, the relative share of primary production in GDP fell to 11 percent while the share of manufacturing quadrupled to 32 percent. This was inevitable because a sizable gap in the average growth rate between agriculture and manufacturing has persisted: 3.5 percent against 14 percent a year. Chenery and Syrquin (1975) found that "in the case of Korea the primary share was substantially below, whereas the industry's was quite above the international norm." Thus, Korea's economic development has been led by industrialization, not agriculture, and the latter's growth has been in no small measure induced by the former.

Why Industrialization Was Pursued Ahead of Agriculture

The process of industrialization in Korea was inevitably led by exports because of the lack of natural resources and the small size of the domestic market. Korea is, like Japan and Taiwan, a typical Lo-Salih Model IV (East Asia) country, characterized by "high population density, low resource endowment, and use of high technology." Korea's population density is 423 persons per square kilometer, and only one fifth of its land is suitable for farming. Its main resource is the quality of its people, owing to education. Given these preconditions, it was imperative for Korea from the outset to adopt an "outward-looking" development strategy.

The extent to which industrialization is linked to exports is indicated by the rapid increase in the export of manufactured goods as a percentage of total production. Manufactured exports constituted only 4 percent of total manufactured output until 1962, but thereafter rose rapidly to reach 22 percent by 1975. In contrast, manufactured imports, which had accounted for nearly 40 percent of domestic demand in 1955, declined to 21 percent by 1975. Overall, Korea's trade balance achieved a surplus for the first time in 1977 and has continued to increase ever since. Chiefly led by manufactured exports, dramatic structural changes have occurred not only in production but also in trade and domestic demand. This is why the endeavor to bring about these structural changes was called the "export-oriented industrialization" strategy.

The outward-looking industrialization strategy, however, inevitably resulted in the relative neglect of agriculture and rural areas. From the 1950s to the 1960s, the agricultural sector suffered from the government's low grain price policy, which favored urban consumers. Public investment in agriculture was also restrained in order to render more resources available to the industrial sector. However, the government did not completely neglect agriculture; Rather, it made efforts to boost farm production as much as possible.

How Industrialization Induced Agricultural Development

The success of initial industrialization in the 1960s enabled the government to allocate more resources for agricultural and rural development during the 1970s. Thus, in late 1969 the policy of pricing rice high in the dual price system was introduced to favor both farmers and urban consumers; it proved to be an effective incentive to enhance grain production. The Saemaul Movement was also launched in 1971 and its main objectives were to increase farm production and income as well as to improve the rural environment. The government provided technical and financial assistance to encourage small income-generating projects on a voluntary and cooperative basis. One of the key indicators of its success was that the gap between urban and rural household incomes almost disappeared by the mid-1970s.

In addition, industrialization lowered the prices of farm inputs like fertilizer and farm machinery, thereby improving agricultural productivity. It also absorbed excess labor from the countryside. Because the total cultivated acreage remained unchanged and the farm labor population declined, the sources of increased productivity were cheaper fertilizer and machinery inputs. Other sources were the effects of rising urban demand for highly valued cash crops like vegetables and fruits. The construction of an extensive road network to link industrial centers also induced the rapid growth of vegetable production and of other cash crops that were destined for the urban markets.

How Agriculture Supported Industrialization

The limited role of agriculture in the overall growth can be seen in other ways as well. Taxation on the rural sector was very light, amounting to no more than 2 or 3 percent of farm household income, and rural savings, if any, remained in the rural areas. In fact, it was not until the 1970s that there was any net cash flow out of rural areas. The main contributions of agriculture to the economy, therefore, were people and food. The problem of food will be discussed in the next section, but it should be noted that the outmigration of rural people to urban areas has been a prime contribution that agriculture has made to industrialization.

The shift of the labor force from the rural to the urban sector characterizes development in any country, but in Korea it has been extremely rapid. The farms have supplied the surplus labor that has fueled industrial expansion. Korea, in fact, ceased to be a labor-surplus economy by the mid-1960s. Real wages in urban areas rapidly rose in around 1965, and the increasing scarcity of labor on the farm was one of the factors responsible for accelerating farm mechanization beginning in the late 1960s and expanding in the 1980s (Table 2).

In sum, Korea's industrialization itself created conditions conducive to agricultural development. It generated an ever-growing and effective demand for agricultural outputs, supplied inexpensive fertilizers and farm machineries, and absorbed excess rural labor. Agriculture in turn assisted industrialization in two main ways: by providing cheap labor and cheap grains.

THE PROBLEM OF FOOD AND AGRICULTURAL SUFFICIENCY

Korea has steadily sought to achieve "adequacy" rather than "self-sufficiency" in its food production and agricultural development. Agricultural sufficiency in Korea has been conceived in relative terms of maximum feasibility, rather than as autarkic sufficiency. The government never strove to achieve self-sufficiency in all grains and food products. This is a consequence of its basic outward-looking development strategy. As a corollary, food sufficiency in Korea has focused on the staple grains of rice and barley, leaving secondary food to be imported.

Overall Status of Korea's Food Sufficiency

Historically, Korea was self-sufficient in food, owing to the predominantly agrarian economy. However, the volume and pattern of demand for food have changed greatly as a result of rapid economic growth. As a consequence, the overall ratio of Korea's food sufficiency has deteriorated remarkably. As Table 3 indicates, Korea enjoys self-sufficiency in staple grains, vegetables and fruits, livestock, and fishery products. On the other hand, other grains are in short

Table 2. Farm Implements and Machines

	1975	1988	Times
Farm population ('000)	13,244	7272	(−45%)
Farm labor ('000)	5339	3484	(−35%)
Power tiller	85,725	25,803	8.5
Farm tractor	564	24,616	43.6
Power planting	0	92,067	
Harvesting	00,000	00,000	1248.1
Threshing	127,105	296,854	2.3

Sources: Ministry of Agriculture, Forestry & Fisheries

Table 3. Ratio of Self-Sufficiency in Food (Unit: %)

	1965	1975	1988
All food	95.5	97.4	85.5
Staple grains*	95.7	101.6	100.4
Other grains	66.8	46.6	3.7
Livestocks**	83.4	100.3	89.7
Vegetables and fruits	99.1	100.8	100.6
Fishery products	114.2	136.8	133.0

Sources: Economic Planning Board and The Bank of Korea
*Covers 1961
**Covers 1971

supply, e.g., the ratio for wheat is 0:1 while that of soybean is 18:4. Korea's track record on grain sufficiency, therefore, is a mixture of success in stable grains and failure in other grains.

In might be interesting to look at the current consumption level and past trend of major food items. As shown in Table 4, all items have greatly increased in terms of annual per capita consumption, reflecting the rising income elasticity of demand for high-grade diet.

Production Policies for Food Adequacy

The Korean government has on the whole been successful in its food production policies. Although the overall ratio of Korea's food sufficiency has deteriorated (85.5 percent in 1988), agricultural labor productivity has been improving at an average annual rate of about 6 percent. As a result, domestic production of major food items has increased (Table 5). Once again this is evidence of the great impact of industrialization and urbanization upon the production of agricultural outputs because they are in increasing demand by urban consumers with rapidly rising incomes.

Table 4. Annual Per Capita Consumption of Major Food

	1965	1975	1988
Rice	121.8	123.6	122.2
Barley	36.8	36.3	2.0
Wheat	13.8	29.5	33.9
Vegetables	45.5	62.5	115.1
Fruits	11.3	15.0	39.9
Meat	4.6	6.5	19.5
Milk	2.1	4.4	40.0
Fishery products	18.0	29.9	39.7

Sources: Ministry of Agriculture, Forestry & Fisheries; Korea Rural Economic Institute

Table 5. Domestic Production or Major Food (Unit: 1000 M/T)

	1965	1988	Times
Rice	3501	5493	1.6
Vegetables	1577	7651	4.9
Fruits	311	1741	5.5
Meat	168	958	5.7
Milk	13	1632	125.5
Fishery products	626	3209	5.1
Farm labor (1000 persons)	4742	3484	.7
Acreage per farm (ha)	0.95	1.18	1.2

Source: Korea Rural Economic Institute

Let me now explain the ways in which these notable gains in food production occurred in spite of Korea's "industrialization first" policy. Aside from the sweeping land reform from 1949 to 1952, it has taken a series of incentive schemes. In 1958, the Agriculture Bank was organized to extend credits to farmers and the Office of Agricultural Extension was established to disseminate modern farming technology. In 1961, the former was reorganized as the National Agricultural Cooperatives with added marketing function, and the latter integrated the Central Bureau of Community Development to become the Office of Rural Development (ORD). A number of production-enhancing projects followed in the 1960s: improving irrigation facilities from 1962, developing slopeland and tideland from 1963, consolidating paddy land from 1964, building five fertilizer plants to achieve self-sufficiency with subsidized sales to farmers, and enacting the Basic Agricultural Law in 1967 to strengthen the legal basis of farm policies. The agricultural sector grew at an average annual rate of 3 percent in the 1950s and 4.5 percent in the 1960s.

In the 1970s, the government intensified its efforts while pursuing further industrialization. Alone with the bold shift in the grain price policy from the low to the high dual price system in 1969, ORD succeeded in 1971 in developing a new rice variety (Tongil) that alone increased yields by 25 to 30 percent. Farm mechanization was also actively carried out from 1972 to 1975 with the public outlay of 3.8 billion won under the generous condition of a 40 percent subsidy and a 60 percent long-term loan. All these rural development efforts were combined in 1971 under the Saemaul movement. The agricultural sector continued to register respectable growth in the 1970s, at an annual average rate of 4 percent. It is deplorable, however, that agricultural growth has been stagnant in the 1980s, largely because of the implementation of structural adjustment policies.

Price Policies for Securing Food Adequacy

Food adequacy in Korea has relied more on price policies than production policies. The low grain price policy continued through the 1950s and 1960s in order to secure adequate supply at prices that would prevent inflation. It must be recalled in this regard that in Korea "grain price policy" has been virtually identified with "farm price policy" because of its dominance in Korean life. This has been particularly true of rice.

Under the low grain price policy, the government acquired grains from farmers at set prices, usually 10 percent to 15 percent below the market price, through direct purchase, rice-fertilizer barter, and taxation in kind. Farmers suffered losses because of the government grain policy, which discouraged grain production while encouraging consumption and thereby widened the food gap, which had to be met by imports. In 1962, the import of grains amounted to $40 million, most of which was aided by US Public Law 480, but the figure increased to $245 million by 1970, and had to be financed by Korea's own foreign exchange holdings. Because of the rising foreign exchange burden added to the dwindling domestic production in the late 1960s, the government was forced to adopt the high grain price policy in late 1969.

The new policy was called a "dual grain price policy" because the government purchased grains from farmers at an incentive price higher than the market price while selling to consumers at a price 15 percent lower than the purchase price. This served the double function of dampening inflation as well as boosting grain production. However, the deficit had to be financed by long-term overdraft from the central bank, thereby causing a large increase in the money supply. For instance, it amounted to 23 billion won in 1975, which accounted for 98 percent of the increase in money supply that year. Some balancing efforts to reverse the trend have been made since 1982, but the accumulated deficit is snowballing. At the end of 1988, it stands at 3.3 trillion won, which amounts to 10.6 percent of total government expenditure.

SOME LESSONS FROM THE KOREAN EXPERIENCE

The old saying "success entails problems" warns us that economic development tends to elicit rising expectation, thereby creating new problems that defy an easy solution. The Korean economy is now on the brink of transition from structural adjustment to maturity, yet it faces some problems, especially in its agricultural sector. I will confine my discussion to agricultural problems and some of the possible lessons to be learned from the Korean experience.

Agricultural Problems and Failures

The most critical structural problems confronting Korean agriculture are (a) the accumulated fiscal deficit caused by food grain policy; (b) the sagging agricultural productivity as evidenced by average annual growth rate of 2 percent since 1978; and (c) the widening income disparity between rural and urban sectors in the 1980s. Probably the major factors responsible for these problems are (a) the direct impact of structural adjustment since 1979, which minimized incentive schemes previously given to the agricultural sector; (b) the virtual saturation of grain outputs as affected by production and price policies; (c) the setback of the Saemaul movement after the death of President Park who was its mastermind; and, most recently, (d) the farmers' resentment against the ever-expanding import liberalization.

It seems to me that the current spate of problems stems fundamentally from Korea's basic development strategy, which has consistently favored industrialization while neglecting agriculture. For example, the government's stress on rice production has resulted in a lack of crop diversification, which today incurs a large foreign exchange burden for imports. The same can be said of its neglect of rural industrialization to provide off-farm incomes and employment opportunities. These problems can be better appreciated when compared to the case of Taiwan. The two countries started from roughly the same condition and both experienced successful land reforms. In the case of Taiwan, however, the government strategy followed the conventional wisdom of stressing agricultural development first. As a result, for example, average farm family income in Taiwan has been about twice that of Korea.

Some Possible Hints

Let me know consider what lessons, if any, one might draw from the Korean experience in agriculture. I am quite skeptical of the possibility that the Korean experience can be reproduced in the African context. In the final analysis, it is our understanding of the local context of constraints and opportunities that must determine the course of development. Development is a complicated process involving various interacting factors, not only economic but also political, social, and cultural. With this precaution in mind, I would like to offer a few hints, not lessons, for consideration.

Perhaps the most significant is the importance of land reform. No doubt many nations have land reform laws on the book but little in practice. Korea undertook what one author called "one of the twentieth century's most thoroughgoing land reforms." Of course, chaos haunted government efforts and yet when the process was completed, oppressed rural Korea was transformed into a viable egalitarian society, which provided the basis for later modernization. In the case of Korea, the direct impact of land reform lay in the area of income redistribution more than on agricultural productivity.

The next but no less important factor is the voluntarism of the rural people. Without a will and voluntary participation in village programs, even the best policy or generous external aid cannot produce positive results. To this effect, the Saemaul movement upgraded the quality of rural life and transformed age-old rural attitudes. It was local self-help effort led by a village volunteer with the underlying spirits of self-help, hard work, and savings. Let me repeat that voluntarism is a key to rural development.

Last but not the least is the linking of agricultural development and industrialization. Agriculture can neither stand by itself nor is its development an end in itself, whatever its magnitude in the economy. In view of its relatively low productivity and static nature, it is better for agriculture to be stimulated by external dynamics like industrial spurt, export drive, or technological input. Because these external factors are tied to the process of industrialization, the Korean model of industrialization-induced agricultural development may prove valuable for other countries, even if Korea has gone too far in the direction of industrialization. Let me stress again that there should be a synergy between agricultural development and industrialization.

Conclusion

It is worth recollecting the hopelessness of Korea a generation ago, when we Korean planners were groping for the first five-year development plan. Korea was then resource-poor, overpopulated, war-torn, poorly fed, and demoralized, not knowing where to start. A British journalist visiting Korea at that time correctly reported: "You can't grow roses in a garbage basket." I hope you can imagine how much we planners were discouraged to hear this. Indeed, no Korean in the early 1960s, including myself, who was fortunate enough to serve as the planning director in the government during 1962 to 1963, could have anticipated such rapid sustained growth reviewed above.

From this personal experience, I do not hesitate to offer one lesson, not just a hint, for the conviction that an economic takeoff and self-sustained growth is "possible" only if the people are ready and self-reliant. As regards self-reliance, it should be noted that Korea's economic takeoff began only after 1963 when economic aid from the United States ebbed. Korea has demonstrated that development is possible! At the very least, the Korean experience indicates that despair over the possibility of economic development is altogether unwarranted.

Development means mobilizing individual talent and energy for the release of the human spirit and its fuel is the people's desire for development. Korea's first five-year plan was nothing but the cumulation of the people's desire for rapid economic growth, deriving from the pressures of poverty, frustration, and despair. Therefore, the question is who will and how to ignite the people's desire.

REFERENCES

Adelman, Irma, ed. *Practical Approaches to Development Planning: Korea's Second Five-Year Plan* (Baltimore: The Johns Hopkins University Press, 1969).

Ban, Sung Whan et. al., *Rural Development* (Cambridge: Harvard University Press, 1980).

Brown, Gilbert T., *Korean Pricing Policies and Economic Development in the 1960s* (Baltimore: The Johns Hopkins University Press, 1975).

Chenery, Hollis, and Moises, Syrquin, *Patterns of Development: 1950–1970* (Oxford: Oxford University Press, 1975).

Cole, David C., and Lyman, Princeton N., *Korean Development: The Interplay of Politics and Economics* (Cambridge: Harvard University Press, 1971).

Hasan, Parvez, *Korea: Problems and Issues in a Rapidly Growing Economy* (Baltimore: The Johns Hopkins University Press, 1976).

Keidel, Albert, III., *Korean Regional Farm Product and Income: 1910–1975* (Seoul: Korea Development Institute, 1979).

Kim, Kwang Suk, and Roemer, Michael, *Growth and Structural Transformation of Korea: 1945–1975* (Cambridge: Harvard University Press, 1979).

Kuznets, Paul W., *Economic Growth and Structure in Korea* (New Haven: Yale University Press, 1977).

Lo, Fu-chen, and Salih, Kamal, *Regional Policy and Rural-Urban Transformation in Asia: Proceedings* (Nagoya, 1977).

Mason, Edward S., et al. *The Economic and Social Modernization of the Republic of Korea* (Cambridge: Harvard University Press, 1980).

Moon, Pal Yong, and Ryu, Byung Seo, *Korea's Agricultural Policies in Historical Perspective,* Korea Development Institute Working Paper 7704 (Seoul: Korea Development Institute, 1977).

Ryu, Byung Seo, and Song, Dae Hee, *Agricultural Policies for Structural Adjustment in the NICs: Lessons from Korea.* Korea Development Institute Working Paper 8611 (Seoul: Korea Development Institute, 1986).

Scalapino, Robert A., "Korean Politics: Tradition and Change in an Era of Crisis," In Changsoo Lee, ed., *Modernization of Korea and Impact of the West* (Los Angeles: EASC, University of Southern California, 1981).

Westphal, Larry E., and Kim, Kwang Suk, *Industrial Policy and Development in Korea.* World Bank Staff Paper No. 263 (Washington, D.C.: World Bank, 1977).

Afterword: Presentation by Jimmy Carter and Discussion

Introduction by Olusegun Obasanjo

I need hardly say that for me it is both an honor and a great pleasure to welcome President and Mrs. Carter to Nigeria once again, and particularly to our town, Ota. When President Carter as the incumbent president of the United States honored us with that epoch-making visit to Nigeria 11 years ago, I did not have to introduce him. So you can understand when I say that I find myself now in something of a dilemma. How does one introduce a man who needs no introduction and still sound relevant?

As I turned it over in my head, I was reminded of a recent experience I had in the United States. At the end of May, I addressed the New York Harvard Club on the rather sleepy subject of the future of African nationalism. The chairman on that occasion was one of those intrepid personalities that only American genes can produce. He presided over the proceedings with exquisite tolerance and a seriousness of purpose appropriate to the occasion. At the end of my presentation, when I agreed to take questions from the floor, he warned those members of the audience intending to ask questions to keep quiet if they could not improve on silence. I have had occasion to be grateful to that chairman for his guidance. But this will not be one of those occasions. On this occasion I am actually confident that I can improve on silence.

President Carter is here in several capacities. Like myself, he is no longer in harness. He comes to us with the incomparable insights gained in office but without the trammels of officialdom. He will therefore be speaking to us with unique authority and his characteristic candor. Like myself and others in this room, President Carter is a farmer. He will therefore be speaking to us from the particular vantage point of a farmer. As you all know, President Carter preceded me in farming by a good many years and achieved a roaring success in it. I cannot claim that success yet. So I have come to look on him as Dante looked on Virgil, as a teacher and a model. But having succumbed to the analogy of this classical parallel, I must now honestly ask you not to compare Ota, Nigeria with Plains, Georgia.

We rejoice in the President's presence today for other reasons. He now devotes a great deal of his time and energy to issues of African development, human rights, and conflict resolution. The Carter Center, which he established on leaving office and which works under his personal oversight, sponsors a variety of programs focusing on Africa and manages a number of projects in the region. Some of you will be familiar with the achievements of the Center on Global 2000 agricultural projects and the remarkable progress that has been achieved in the fight against guinea worm and river blindness. These are all the direct results of the President's efforts. Now he is launching the Center's latest African program, which involves strategies for food security. A group of experts on African food security has just met in Harare with President Carter. Needless to say we wait to hear the outcome of that meeting with great keenness.

Over the last three days we have gathered here, participants from over 15 African countries and some 6 non-African countries, to deliberate on the same serious issue of agricultural production and food security in Africa. It is an issue of life and death for Africa. Happily, I can say that if we started four days ago with some feelings of despair we are ending today with some renewed hope in the future of Africa to reverse the downward trend in the fortunes of recent times. We have come to the conclusion that what we need is purposefulness, determination, commitment, ambitious and visionary leadership, and supportive friends. Africa knows that it has a reliable and understanding friend. I can assure you that there is a new awareness and a new resolve among Africans, the leaders and the led, to reverse the trend of marginalization of Africa that is observable all around. The Africa Leadership Forum stands ready to continue to be a catalyst in the process of that reversal. The Carter Center and the Africa Leadership Forum are clearly working to the same end in this regard. Cooperation and collaboration between our two organizations should therefore be a matter of course. The modalities for such cooperation also need to be worked out carefully. We in the ALF want to know how best we can contribute to the successful implementation of Project Africa and Global 2000. We are only too aware of our good fortune in having Mr. Carter here to deliver the closing speech of our conference.

Presentation by Jimmy Carter

I have great admiration for and appreciation of General Obasanjo who has been not only a political collaborator in dealing with world events but also a personal friend. On my first visit to Africa, when I was president, it was no accident that I came to Lagos to meet General Obasanjo who was then president of Nigeria.

We shared common commitments, not only in alleviating the suffering of people in this continent, but also to do away if possible with the horrible plight of apartheid and racism in South Africa.

As a fellow farmer, I also felt a common purpose with him, although he has far surpassed my accomplishments in agriculture. I used to have a few sows, maybe 10 at the most, but he has 720 sows. I stand in awe of his accomplishments, political and military, as well as his global commitment to improving the lives of peoples in his own nation and around the world.

I wish I could have attended the law few days of sessions here. I did read the wonderful presentation of Dr. Swaminathan, who is recognized as a world leader in the quest for global food security. The president of our center in Atlanta, Georgia, and the president of our university are both here. In the last few years I have learned so much by meeting with leaders like all of you and seeing at first hand in Africa and Asia what could be accomplished by adequate yet simple farm production techniques.

In 1984–1985, under the general direction of Dr. Norman Borlaug, who is also here with us, we began to analyze what might be done in Africa to correct a situation of increasing concern: the horrible prospect of sustained starvation in nations like Ethiopia or Sudan. We see this tragedy every night on the TV screens of my country and of Europe and around the world. Only when these terrible images were broadcast did the world harness its resources to send food to alleviate the starvation. And this dramatized a problem that existed long before this crisis and continues to the present, an inherent problem in the continent of Africa that has not yet been addressed adequately.

In the last years, because of the sustained efforts of many agencies, many governments, and many people, food production has doubled in Africa. But population is doubling every 20 years! As a result, for the last 20 years, and perhaps even longer, per capita food production in Africa has been going down. To no other region on earth does this problem apply in such a dramatic fashion, and up until now we have not been able to reverse that trend. With a 3.2 percent average population increase, Africa would have to increase food production by 3.2 percent to stay even. But even among those nations most advanced in agriculture, very few have been able to sustain that rate of growth. Something dramatic must be done to reverse this trend and to give the people of Africa a chance to live a life sustained by adequate calorie intake on a daily basis.

What we were able to do is to learn from Dr. Norman Borlaug, who won the Nobel Peace Prize in 1970 for his dramatic accomplishment in bringing about the Green Revolution in India and Pakistan. As you know both these nations were facing horrible starvation in the early 1960s and it was because of the wisdom of Dr. Borlaug and the people he brought with him that now Pakistan and India have become net exporters of basic food grains. At the Carter Center we set about working with financiers from Japan and other countries to see what might be done to bring about a green revolution in sub-Saharan Africa.

We have learned a great deal in the last few years. After a great deal of assessment, we finally decided to adopt four Africa nations—two north of the equator and two south of the equator: Ghana and Sudan, Tanzania and Zambia—in which we might run some test plots to see what might be accomplished. We made our first trip to Africa for this purpose in January 1986. Out of that trip we learned a few things and decided on some basic principles. We went to the four countries and met with the leaders there. The overriding truism that cannot be avoided is that no nation is ever going to substantially increase food grain production without the full and enthusiastic involvement of the King, President, Prime Minister, Minister of Agriculture, and Ministers of Finance, Transportation, Education, and Health. You cannot do it just with a Minister of Agriculture alone; you cannot do it with the most superb scientists and agronomists and extension workers all of whom are dealing with subordinate officials on an isolated rural basis. They can make progress, but not enough to sustain growth and improve production.

What we do habitually is take all necessary factors that go into increasing food production, as learned in India and Pakistan and elsewhere, and put them in a presentation that the simplest farm family can understand and that can be monitored. It is not adequate to have a few square meters of an improved production demonstration; you must have an area large enough so that the farmer will understand it and not have to go through a mathematical formula to calculate how much he would have made on an acre or hectare. Dr. Borlaug has prescribed that the test be at least a half hectare. We choose a few farmers, scattered about and distant from one another, and they plant a right seed—the best seed available for that particular locality—and a moderate but adequate amount of fertilizer. Working with extension workers native to that country we urge these farmers, who are generally competent and admired by their neighbors, to plant at the right time and to carry out adequate weed control. We carefully avoid any dependence on mechanization. The farmers sometimes plant with a pointed stick and cultivate with a hoe. We have been able to accomplish impressive, amazing results. In Zambia, Sudan, and Ghana Dr. Borlaug's people have been able to bring about a routine doubling or tripling of sorghum, maize, wheat, and millet. In some cases, increases have been five- or sixfold.

Yesterday, in Zambia, Dr. Borlaug and I, along with some other people, visited a farm and met with hundreds of people who came there. The farmer told me that he had always produced about 10 or 15 bags of corn or maize per hectare. He had just finished harvesting 60 bags of maize per hectare. This is a common occurrence. When that test farmer produces two or three times as much as he has ever produced before, then neighboring farmers see this. And that is just what happened in Ghana with 40 farmers the first year. The second year we had more than 1200. Last year, the third year, we had 16,000 farmers and this year we have 85,000 farmers in the Global 2000 Program.

We depend almost exclusively on the local extension workers to carry out the work. Flight Lieutenant Jerry Rawlings, the Head of State of Ghana, and all of his ministers have been thoroughly supportive. Regardless whether I am able to visit Ghana on a trip to Africa, Flight Lieutenant Rawlings is out in the fields when the farmers plant or harvest, publicizing what might be done with the Global 2000 Program. His interest and leadership have made Ghana so successful. Besides the 85,000 farmers in our program, another 200,000 are applying our techniques. Global 2000 has only three people in Ghana and two people in Zambia where several thousand farmers participate. We depend heavily on the leadership and the extension workers in a country. Our hope is that within five years these nations will be fully self-supporting in food grain production and will have developed adequate storage, transportation, and the sustained export market. Then we will be able to leave. But we will leave behind people native to that country who are thoroughly acquainted with the process.

As you can no doubt imagine, we have had problems though in starting these programs. The most serious have been in the Sudan. As you know, that nation has been divided by constant warfare and rapid changes of government. Over the past 15 months we have had to deal with six ministers of agriculture. Dr. Borlaug or I will make arrangements with the minister of agriculture and then leave the country only to find that several months later there is a different minister of agriculture in place. To try to minimize misunderstandings we have been very insistent in requiring that there be a firm contract between the nation involved and the Carter Center. We sign a memorandum of understanding—I and President Mwinyi of Tanzania, former Prime Minister Al Mahdi of Sudan, Flight Lieutenant Jerry Rawlings of Ghana, and President Kaunda of Zambia. We agree to provide certain services and the government agrees to provide certain services.

Another very important aspect of our success has been sufficient credit for farmers to purchase the right kind of seed and an adequate amount of fertilizer. Because there was some doubt about the success of the program, at the beginning we were helped along by a private lending agency—the Bank of Credit and Commerce International (BCCI). They have branches in 72 nations and have been willing to provide very small loans to farmers in the first few years to demonstrate that the process is practicable. In Ghana every farmer repaid the loan in the second year. We have had a 98-percent loan repayment rate because if a farmer can increase from 10 bags of grain per hectare to 60 bags it only takes two or three of those bags to repay the loan. In both Ghana and Zambia the government is now taking more of a role in providing the very modest cost of the seed and fertilizer. After the third and fourth sessions we have learned a lot about how you can practically increase food grain production. We have also learned that you cannot separate food production from the general health of the people, from nutrition, education, transportation, and peace. We have had problems in Sudan and still are having problems, but we are making good progress. It would be impossible for us to have a sustained program in Somalia, Ethiopia, or Mozambique because those nations are devoting their resources to waging war.

Another very important and inseparable characteristic of these problems is indigenous disease. In most farming regions of many nations this affliction is twofold: guinea worm and river blindness. Last year in Nigeria alone 680,000 cases of guinea worm were reported. In one small region of rice production it was estimated that guinea worm reduced the area's income from rice by $20 million in a single year! Guinea worm is a disease that especially afflicts small, remote villages seldom visited by the leader of a nation or by a minister of health. In some countries even where guinea worm is widely prevalent, the minister of health has never heard of the disease. Nigeria is fortunate to have a minister who is an expert in this field, Dr. Ransome Kuti. We have taken on the responsibility of trying to reduce the incidence of guinea worm and in just this past year, working with WHO, UNICEF, UNDP, and others, we have succeeded in having guinea worm targeted for eradication from the face of the earth. This is only the second disease to be so designated, the first having been smallpox over 15 years ago.

The Director of the Carter Center is Dr. Bill Foege, the former director of the Center for Disease Control in Atlanta, who orchestrated the eradication of smallpox. With the targeting of guinea worm, I think we have seen the tie-in of peace, preventive health care, and food production.

River blindness occurs in 30 nations. It afflicts the people who live beside rapidly flowing streams, where a small fly breeds. Blindness is the result of that fly's sting. Quite often the families move away from the streams and abandon what is often the most productive farm land in that country. Thanks to the generosity of an American company a tablet called Impermectin is now being distributed free. Dr. Foege is in charge of this program and any nation that has incidents of river blindness may apply. We will provide tablets free of charge. Only one need be taken once a year to prevent the disease. The point of this is that nutrition, disease prevention, and food production all tie together.

Two days ago, in Harare, we brought together people to discuss the future of what we call Project Africa. We are now readying to expand what we have done and to build upon what others have done successfully in Africa, with the goal to rapidly increase food production. Let me make one point. Many of them know about it but do not assign it high priority on their agenda for action or financing. The Secretary-General of the United Nations, the President of the World Bank, and others of this level of authority must be brought into the process of providing adequate resources for accomplishing food security in Africa.

I want to repeat another point, which I have already mentioned. The heads of state of affected nations must be directly and personally involved. Without their sustained support and interest, almost any program we implement is not going to be continued. We need a thorough assessment of the successes that have been realized to convince people that the situation is not hopeless. Demonstrated successes make it easier to convince discouraged leaders and citizens that something can be done about our problem. Each agency must be given freedom to pursue its own interest and to expand its own successful programs without domination from any other agency. The economic policies of a country must be modified. Farmers cannot be induced to produce more grain without fair remuneration, otherwise they move then to the cities. Farmers then lose the political influence they formerly enjoyed. In economic crisis, heads of state are inclined to reduce the price of grain to satisfy city dwellers below the cost of production for farmers. Last year, about 25 percent of the total wheat production was lost in Sudan because at the last minute the government reduced the price so low that the farmers refused to harvest their crop. I went to Sudan and talked to Sadiq Al Mahdi, then prime minister, and encouraged him and his ministers to raise the price of wheat so that the crop would be harvested. It worked. So, economic policies must be included.

There must be a closer correlation of research findings and application in the fields. Research scientists often are not particularly interested in the wide use of the superior seeds or techniques that they develop. If they produce a seed—sorghum or maize—that can produce three times as much as indigenous seed, they are not so much interested in getting that seed cultivated all over the country as in developing a seed that will yield four times as much as the indigenous one. Meanwhile, decades may pass without improved seed being used. As well, the government must be willing to allow private seed producers into the country. Dr. Borlaug frequently points out that parastatal production alone of hybrids or new varieties, as is the case in some countries, is often a very inefficient operation. Seeds cannot slip through the bureaucracy to reach the farmers. There must be an awareness of how we get adequate fertilizer and adequate seed of improved varieties. The grain research centers in Mexico and India need to have the product of their research actually applied to individual small farmers throughout Africa.

Another point that has to be understood is that you cannot improve food production in countries unless you deal at the small-farmer level. The establishment of major corporate farms or state farms will not solve the problem. The small farmer has to be addressed individually. In most cases, women manage the farm and make the decisions about food production. To ignore the female leadership that exists at the local village level is a serious mistake. There has to be maximum involvement of local workers so that when the scientists leave or when major programs change, the indigenous farmers, ministers of agriculture, and extension workers can continue.

I think we need to cooperate with each other globally to say where fertilizers come from, and where we can acquire adequate seed. Some nations have a burning natural gas that could provide urea or nitrogen for the entire continent. The building of plants to produce nitrogen and the acquisition of phosphate or potash are something we need to address collectively. There are times when the world community needs to concentrate on preventing problems that persist. Transportation, economic policies, war, adequate storage, and marketing capabilities—these are the kinds of things that could be studied together, in individual nations and even communities, and dealt with effectively.

Finally, the major benevolent aid agencies and nations need to shift their emphasis from the emergency supply of grains (and of wheat in particular) to applying the same level of financial aid and resources to food production on a sustained basis in a country chronically afflicted with famine. It is not a good program to dump hundreds of thousands or even millions of tons of wheat into a country. That not only distorts the economy of that nation but farmers who are still trying to produce millet or sorghum cannot compete with heavily subsidized wheat. They abandon their farms, move to the cities, and join the ranks of agricultural refugees who are no longer productive. The injection of wheat as a major and very attractive grain into a country that does not have the capability of producing it distorts the program. This is enormously costly. But we could take just the emergency money that has gone into Ethiopia or Sudan in the last few years and bring about an agricultural revolution that would guarantee food security in Africa.

I am new at this subject; many of you have had far more years and much more experience and effectiveness than have I. But the point that can come out of this leadership conference is that working together we can learn from one another and with a sustained effort involving the highest levels of officials and the smallest farmers on the planet, together we can bring about food security and a better life for the people of this continent.

Question-and-Answer Period

Olusegun Obasanjo:

When you started your speech you regretted your inability to be with us for the past four days. From the loudness of the applause that followed the end of your speech, there is some indication that what we have been saying is not dissimilar from what you have said to us. By fluke, I happen to have been able to collect some good minds here from Africa and outside Africa. Those who say great minds think alike must be right. Before you ask questions I want to say this on behalf of our women participants. The remark you made about women, and women farmers in particular, is so apposite and so in line with what women participants have been saying. I think seriously that they have felt that they are not getting a fair deal and some of us feel that they are not getting a fair deal. Getting support from an eminent personality like you will go a long way to reduce their frustration and to make us men realize their frustration. You will realize, Mr. President, that male chauvinism pervades the whole of Africa. Maybe we will move away from our chauvinistic views and actions and give our women a fairer deal within the economic milieu of our nations in Africa.

Question:

In the context of Africa, where producer prices are still low, do you think a lower price for seeds is feasible? I would prefer that the seeds are produced by the existing research centers and multiplied by large-scale farmers.

Is it feasible, in your view, to organize the African youth, many of whom are college graduates and are jobless? Is it possible to organize them to contribute to rural development in the way in which you have organized your Peace Corps from America?

Jimmy Carter:

The cost of good seed is the best investment and cheapest acquisition that a farmer can make because he can plant the seed of his grandfather and father still raise only a minimal crop even though he uses fertilizer. But he can plant seed and make three, four, or five times as much grain with the same work on the same land with the same fertilizer. I don't think that the price of any seed that I know about in the world, if it is good seed and the right kind of variety, is excessive. Dr. Borlaug's experience, and I quote what he told us at the Conference in Harare two days ago, is that when the national governments insist upon maintaining a monopoly in the production of new varieties or hybrids, the quality of seed and the rate of production is not adequate. Sometimes it has been very advisable, as in the case of Zambia, to bring in on a partnership basis private seed producers who are quite competent, and who have proven their promise in other parts of the world, strictly on a minority ownership basis for the seed production agency. I think that is a primary cause and I would not worry about the cost of good seed, because as you know in the case of groundnuts or corn you have a hundredfold increase where corn is planted or where peanuts are planted.

The youth issue is very important. I was in southern Sudan when they announced that we would have a few hours of instruction to look at a demonstration plot and describe to the surrounding farmers what would occur. They expected 20 or 30 people to come; 2000 people came. Some of them travelled three or four days because they had to walk or ride donkeys. Instead of staying a couple of hours they stayed all day because they wanted to know all about it. These were young people, men and women. I think to involve the young generation in this process is extremely fruitful for them. The farmer with whom I met yesterday, near Lusaka, said he had been producing about 10 or 15 bags a day. He had two sons who would like to be farmers. They had moved to Lusaka, but they couldn't get good jobs. When he had his bountiful harvest under the Global 2000 program his sons said they wanted to move back to the farm and join with him, and expand the production from two hectares to six hectares. I think this is the kind of thing we can do with the up and coming generations, i.e., involve them in the process. I don't think there is any doubt that we can accomplish a wide range of goals. Better seed, adequate fertilizer, involvement of youth have a common purpose.

Question:

Where was the research done for the seeds beings distributed to the farmers so that you could move in in 1986 and start getting such good results? Regarding the strategy with BCCI, how are they involved? What happened with regard to interest rates? You were not asking a government to bring down interest rates because of the farmers. You were dealing with a commercial bank and saying this is a good investment. Therefore, their willingness must mean that it paid them within the money market situation. But also regarding the recovery of that loan, I was fascinated. From my own experience, the farmers when properly enlightened about what credit is all about are not the types who would refuse to pay back.

In your concluding remark you highlighted the importance of involving women. We try to do it but in the African context there are all the cultural and religious constraints that make moving women actively into the position they deserve difficult. In the countries you are involved in, Sudan in particular has a strong religious position with regard to women, I would be interested to know how Global 2000 dealt with that problem and how they gave women in those countries the importance they deserved.

Jimmy Carter:

We deal exclusively with maize, millet, sorghum, and wheat. The work, as you know, for these crops is done either at the research center in Mexico or in Indian (ICRISAT). I understand that there are 12 other research centers around the world that deal with rice, cassava and bananas, plantains, chick peas, etc. These research centers have been financed on a global basis adequately. They have emerged as superb organizations. Quite often the product of their work is not delivered to the small farmer who wants to increase his production. This is partly due to an inherent division between research scientists on the one hand and extension workers on the other. Sometimes the research scientists are a little arrogant. They are more inclined to work in the laboratory in tiny plots and they think that the extension workers are not worthy to be treated as equals. Sometimes because of educational differences and the inadequacies of Ministries of Agriculture, this chasm is not closed. The key to Dr. Borlaug's success is that he is both. He is one of the great research scientists of all time; he also spends his time in the field. So we try to mandate that this be the case. When we signed our first contract with Sadiq Al Mahdi's predecessor in the military government, we signed a contract in the name of ICRISAT and involved them in the process. Dr. Borlaug has been able to demonstrate to the leaders of some of these research centers that they ought to do what he has done for Mexico in the case of maize. That is where the research is done and it is adequate. These research centers do not restrict their research to their own country; they have gone for many years into the regions of Africa and developed hybrids that are suitable for those local communities.

Question:

Where does the seed come from?

Jimmy Carter:

It has come sometimes from government agencies, sometimes from seed-producing centers that are controlled in the private free enterprise sector alone, sometimes from a combination, as in Zambia where the government owns 60 percent, and a private seed-producing company owns 40 percent. In some cases that is just not done. Then we have to import the seed. But they ought to be produced locally. With a variety you don't have problems because the farmer can plant the same improved seed a second and third year by saving some seed. With a hybrid he needs to get new seed each year. But this is a very small cost to the farmer. BCCI is unique among all of the banks that I have ever dealt with. They are major sponsors and financiers of the infrastructure of the Global 2000 program. They provide the cost of the scientists, they buy a few pickup trucks and a few more motor scooters, and a lot of bicycles on which our extension workers ride around. In addition, they have seen very clearly that the farmers have to have some way to buy the seed and fertilizer. We don't give away the seed and fertilizer. I think that is the wrong approach. If the farmer thinks he is going to get free seed, free fertilizer, free extension service work, it makes him overly dependent on external sources. We make sure that the farmer understands that he has to repay these loans and we have had markable success—98 percent repayment. When I left Lusaka yesterday, the BCCI chairman was on the plane with me and the local BCCI representative told me that they had already authorized 10,000 farm loans for this coming season in Zambia alone. We demonstrate through BCCI and other private sources that the farmers will repay the loans and we encourage the governments to organize cooperative local banks that will make these loans, with the assurance that if the farmer will adopt the improved production practices it is almost certain that the farmer will be able to repay the loan. This is a very important point.

Regarding women in the Sudan. The first year we started our program, in millet and sorghum and later on wheat, we had 125 extension workers. Forty percent of these were women and they were accepted. Our scientists told me that they were the most effective overall because they knew what was going on. The fact is that you don't have to do anything in Africa to make sure the women make the basic decision to do their part on the farm or at the village level. It is when you get to the regional level with marketing and the government-level decisions that the situation breaks down. Three years ago I went to Harare. We were thinking about a Global 2000 project there, which ultimately did not happen. I went about 100 miles east of Harare in a car to a small farm, because the farmer there had received the "Outstanding Farmer" aware two years in a row and I wanted to meet him. It was a beautiful little farm, the farmer was dressed up in a black suit with a tie on, he had his awards nailed to two trees, and the whole village was there. We went into the home to eat lunch. I talked to the farmer, and his wife was serving food and taking care of everybody. Finally, when I got through eating lunch I said I wanted to go out and see his fields. The farmer began making excuses and said, "Mr. President, you will get your shoes dirty," etc. I said I had been a farmer all my life and wanted to go out and see the fields. He finally, reluctantly, agreed. He walked along with me and my wife and the farmer's wife walked along behind us. We got to the maize field. It was a beautiful field of maize. I said, "Is this a variety of hybrid?" There was a long silence and he turned to his wife and he said, "Is this a variety or a hybrid?" I asked, "How much fertilizer do you use per acre?" He turned to his wife and asked her. She said, "We use 600 pounds per acre." I asked him about herbicide—he didn't know, he didn't know about the planting date, he didn't know about the cattle. The point is that in the case of this outstanding farmer who received the awards for two years, his wife did all the work and made all the decisions. At the family level the women are making all the decisions. It is giving them credit for what they are accomplishing, involving them at a higher level in the agricultural hierarchy that is the problem.

Question:

Since you are already registering such success in the four countries, when will you start testing in other smaller countries like Benin? You talked about the number of extension workers and I was very much impressed that you concentrate on using the local resources. In Benin, for example, we have European volunteers for whom one volunteer organization spends about $3000 per month, whereas the average salary of those working is about $50. No person from the university is going to get a job in the next four years. How would Global 2000 help to develop this tremendous human resource that is being wasted today?

Jimmy Carter:

Because of the high publicity of our program in Ghana, we have been approached by the president of Benin. We have a tiny program in Benin already. I think 85 farmers this year have participated in our Global 2000 program. Our three people in Ghana handling 85,000 farmers have been going over for just a few days to Benin to get the program started. We don't yet have the funding or personnel resources to go to other nations. We have a very tiny entrée also in Togo. But we have concentrated on the four nations to learn what I have described to you. Dr. Borlaug has told me that we are looking favorably on going into Benin. How do you get the extension workers? Here again, there is a hidden treasure in your countries. Over the last 20 or 30 years under the requirements of USAID, FAO, and the World Bank, etc., the research centers operate a training program for young native men and women. They come to the research centers such as ICRISAT and CIMMYET and undergo a training of three to six months and go back as qualified extension workers. Now they are working in the post office or they are teaching primary school or trying to get a job in service station or ironing clothes. We have the records of these extension worker as they went through the training program and when we sign a contract with the head of a nation we include in that the commitment from the head of the nation to provide the people we want to be our extension workers. So we pick out the young people who have come back to Ghana, Tanzania, Sudan, etc., who did the best job at the research centers and we ask for these people to be our extension workers. They are already trained. In all of your countries you have had to participate in this research training program and to identify these young people and put them back into agriculture. This is something that is available to you. This is something you can do when you get back home—inventory the list of people who have been to these six research centers and been educated already and have come back but are not working in agriculture. I am sure they would like to have a nice, reasonably paid job in the profession they have chosen.

Dr. Norman Borlaug:

Nigeria has all records of people who have been trained in these centers. They should be recruited and put to work in the field in which they are trained and competent. It does not take much stimulus to get these people strummed up again so that they are enthusiastic. We have no vast sums of money in our program but to make an extension worker functional we buy them a bicycle, sometimes a pair of rubber boots if they are wading in the mud. They pay for those bicycles over two or three years' time out of their own pocket and you would be surprised at the enthusiasm and effectiveness of those workers.

Jimmy Carter:

The point that Dr. Borlaug brought out is the low cost of our program. In all four of these nations with 85,000 farmers in Ghana, 30,000 farmers in Zambia and so forth, the cost is very low. The first year we start out with about $500,000 and the second year about $1 million and then on a sustained basis about $2 million per nation. That is all we put into it. The native government pays the salaries of the extension workers; we don't do that. The financing of seed and fertilizer is paid for by the farmers; we don't pay for that. We provide a few pickup trucks—we just gave the people in Zambia three just to get around. We also provide a few motor scooters for supervisors and a lot of bicycles for extension workers. The extension workers pay for the bicycles out of their salary. It is a grass-roots thing that is not very costly but requires cooperation from the top to the bottom.

Cheikh Sarr:

Some years ago people were criticizing the international donors for not having a harmonized approach. That is why we created consultative group meetings, international donors' meetings, etc. I know your initiative is a good one. Some other people may have other good initiatives. How do you see the relation and the need to harmonize all these actions, to avoid the Ministers of Agriculture having to speak to many different people about many different approaches?

In Africa it is very difficult to finance agriculture or social sectors (education and health) by other funds than concessional funds. As you know, the regional banks as well as IDA and IFAD and the AFDB replenish their soft windows every three years and it is replenished with funds from developed countries.

Every three years the negotiation takes a lot of time to convince them to give more funds. The needs are identified and they are convinced but to get them to give more funds is very difficult. How do you see the role of personalities like you and General Obasanjo to lobby or convince those people in power to give more funds to the regional banks?

Jimmy Carter:

The reason for our meeting in Harare was to bring together the AFDB, World Bank, UNDP, and other agencies to work out a harmonious, cooperative, and mutually supportive approach. So far, every penny to finance the work done by Global 2000 has come from private contributors—from the Sasakawa Foundation in Japan and from BCCI. Our entire guinea worm program in Pakistan and other countries has come from BCCI. We are meeting tomorrow in Lagos to marshal a sustained approach to eradicating guinea worm and we have targeted the year 1995 for guinea worm's eradication from the world. I don't think there is any doubt that we can succeed. But we have limited financial resources from one small private bank, BCCI. They have supported us so far but we need to involve other people. We cannot expand our programs to other nations without full support from you and others. How do you harmonize? I think this kind of meeting is an excellent way to harmonize because I don't think there is any doubt that when you and I leave here we will have learned a lot from one another. We have seen that Global 2000 and AFDB can contribute their part, but if we work together we can greatly magnify our individual efforts. I can contribute my access to world leaders. There is no problem for me to get an appointment with the Secretary-General of the United Nations. If I call him this afternoon he will tell me to come to his apartment tonight for a private meeting. I do this when I want to, with the president of the World Bank or President Babangida or Flight Lieutenant Rawlings. I met this week with President Mengistu in Ethiopia and with the new leader in Sudan. I have no problem with this. I find when I go there they are surprised that I want to talk about guinea worm. I can relate to them from what you teach me, I can relate to the Secretary-General of the United Nations that he should concentrate on food security in Africa. It is very important [because] it affects agriculture, nutrition, survival, birth control, war, and peace. You can't separate them. So this is a very important element to realize.

The purpose of our meeting tomorrow is to expand the financial base on which we have to predicate the eradication of guinea worm. I wish all of you could go to a village where guinea worm is prevalent. Most of you probably never heard of guinea worm. I have been to nations where the head of state and even the minister of health never heard of guinea worm and they may have hundreds of thousands of cases of guinea worm that afflict their country. It comes exclusively from drinking filthy water within which there are guinea worm eggs. These go into the body and 12 months later the egg grows into a long worm and emerges from the body by penetrating the moist places of the skin from the inside. You can imagine where the moist places are and the person is totally incapacitated. The worms emerge during planting season as the weather warms up. It is a devastating blow to the production of food grains. I think that not only the minister of health but also the minister of agriculture should be leading the drive to do away with guinea worm. But there is no doubt in my mind that by 1995, if we work together, we can totally eliminate this disease. Once you break the cycle for 12 months, guinea worm, which only goes into humans, is eradicated from a village unless somebody brings it in from outside.

Concerning river blindness, two years ago Merck & Company came to the Carter Center and talked to Dr William Foege. They said that if we would develop the procedure for the distribution of Invermectin (Mectozan) we will provide this medicine free of charge for every nation on earth in perpetuity. If a villager takes one tablet a year, he will not have river blindness. We have now signed up 20 nations to undertake the prevention of river blindness. It may be possible to eradicate this in the future but we have not addressed that yet. We just want to control at a local village level the incidence of this blindness. You can't separate increased production of rice or sorghum or whatever in a small village that has river blindness. How are they going to plant and cultivate the crop, harvest and market it? They can't do it.

Mrs. Adeni:

I would like to thank you on behalf of the women for accepting the importance of women in agriculture. We have been talking about grains and health a lot. But we know that we need some animal protein. I would like to ask about livestock, i.e., poultry, sheep, goats, etc., because when we produce the grain, human beings can't eat all the grains. We need enough animal protein so that our own health will improve.

Jimmy Carter:

I have to remind you that we only started in agriculture in Africa three years ago. We are likely to confine the Global 2000 emphasis to grain. One of the additional programs we have at the Carter Center is to try to bring about peace within countries that have revolutionary or civil wars. I don't think there is any way to revive the economy of Sudan until the war is over. As you know, the Dinka people have been the main ones to suffer. They, like the Masai, depend almost exclusively on cattle. That is a very worthy project to undertake, i.e., the restoration of the lost source of livelihood and the improvement of breeding and grain for cattle. But I don't want to mislead you; I don't think the Global 2000 program will ever adopt cattle or livestock as a means for alleviating the food problem in Africa. I don't want to discourage anybody but this is one of the causes of our problem. It takes from seven to ten times more grain to provide a certain level of food protein for human consumption through an animal than if you eat the grain yourself. I am hesitant to say that on General Obasanjo's farm, which is producing such wonderful food for his people. But if a country is literally starving then the investment should be made in grain production. I have been very seriously grieved at what has happened in Egypt. It used to be self-sufficient in food grain production; now the biggest import item is food grain. The heavy foreign debt, which is such a burden to the Egyptian people, exists because they are importing so much wheat. The farmers can't make a profit producing food grain any more because they have to compete with imported wheat at a very highly subsidized price. What is happening in Egypt is that the farmers are taking imported wheat, making bread, feeding it to goats, and selling the goats for profit. You have a problem here that is changing, disrupting, or modifying the basic agricultural economy of a country because of a basic dependence on imported wheat. I think you have to balance this. Obviously you can't change the lifestyles of Masai or Dinka and say you can't live on cattle anymore. They are going to do this for the next 10,000 years. It is important to increase the production, quality, and efficiency of animals but I don't think Global 2000 will get involved in this. Other people will have to take on the responsibility of livestock.

Question:

I want to join in thanking you for giving recognition to the place of women in agriculture. One of the causes of the problems we had in the past for the failure of agricultural programs to take off is the exclusion of women from the programs. It is heartening that people like you are giving recognition to this aspect of labor and planners in agriculture. I am sure you know that the incidence of diseases, especially of guinea worm and river blindness, is the greatest in women. Because of this past negligence to women's contribution to agriculture, women have become highly marginalized. It is not enough now to recognize women and include them in the program; you need to do more. Start them at a level so that as time goes on they will be able to catch up. I am suggesting that your organization and ALF see to this. Women should not just be put into the mainstream; if you do this there is no way of catching up. There is no way of our being able to make the necessary contribution required from us. The special arrangement for women should take into consideration the fact that they have been suffering from diseases more than men.

Eugene Terry:

There are 13 centers, as the West African Rice Development Association recently became a fully fledged member of the CGIAR. I was particularly pleased when you talked about the human health hazards and what your organization is doing to tackle these problems. My business is rice. Even though the majority of the people who grow rice grow it on the uplands, the best place is wherever the water is. The environment is more robust and we do not have the problems of sustainability if we put water and rice. In fact, that is why the Asians are able to produce much better rice than we have been able to do. However, one of the constraints of moving the rice producer down to the swamps to the water is the human health hazards. There are three or four very severe diseases that constrain the production of rice in the water parts. Recently, when we made our presentation to the CGIAR we were told that we have to incorporate the human health aspect of research into our work if we want to make any progress in moving rice down to the wetter parts.

My question is: what plans do you have about harmonization, what plans do you have to see if we can link funds for human health research with agricultural development and research? Recently we were told that it is the environmentalists who are making it possible for us to sustain the level of support we are getting for agricultural research. I am very sure that very soon it is the health people who will help us to sustain the amount of money we get for research. I see a unique opportunity here to marry some of these aspects of our work.

William Foege:

I can say that our major problem is to apply what we already know. We have a vast amount of information on health. I think it is very rewarding to see how much health has improved in Africa and this is not totally recognized around the world. In the last 25 years the life expectancy in Africa has been increasing twice as fast as the life expectancy in the United States. This means that the gap is closing in health. At the same time we can't be too pleased until the gap is totally closed. I think the first step in closing that gap is to apply the information already available in guinea worm, in river blindness, and in child immunization. But that should not stop us from investing in the new research that is needed in the future. So more emphasis is needed on application but we still need to continue doing research.

Jimmy Carter:

When we make a step forward we quite often face an additional obstacle and the recent knowledge in my country and in many countries in Africa about AIDS is one of the most sobering things that we face. The Center for Disease Control and others are trying to find a vaccine for AIDS. It has not yet been found. We have visited some countries on this trip where 25 percent of all the newborn babies have AIDS, which they cannot survive. It is probable that before they are seven years old all these babies will be dead. Among some groups of adults the incidence of AIDS is even higher, such as in army troops that have been in regions where the prostitutes are infected with AIDS and where as many as 40 percent of these soldiers test positive for AIDS. These are problems for the future.

I think as we learn how to address the problem of guinea worm, we will develop a health infrastructure or system that will prove to be effective to deal with currently known problems and also to deal with those that come along in the future. This is a very important lesson. Nigeria will be the major test case in the success of eradicating guinea worm.

In some countries there is a direct relationship between the federal or national government on the one hand and the local governments on the other. In

Nigeria you have the national government, 21 state governments—a combination of the military leaders who are the governors on the one hand and civilians who are ministers on the other—then you have the local government. No matter how hard we try, this is a problem that creates a bureaucracy that might be an obstacle rather than an asset to progress. One of the things we hope to accomplish here by this very brief visit to Nigeria is to help bring about a direct relationship between experts like Dr. Ransome-Kuti and others in his Ministry of Health and the local leaders out in the field and not have to go through the cumbersome bureaucracy of a complicated state government. Quite often at the state government level the governors change, they go back to their military duties and you get another governor, whereas at the village level every year the people who suffer from guinea worm are seen. I think this is a very important aspect when dealing with agriculture or health, i.e., to have a free, easy, and unimpeded flow of information, education, training, financing, and administration between the people who are convinced at the highest level on the one hand and those who actually carry out the programs and deliver the services at the local level. This decision cannot be made in the UN, in Washington, in Europe, or Japan. It can't be made at the local level in your countries. It has to be made by the top leaders who can exert their authority and say this is the way it's going to be.

It is only when you can bring together the head of a nation, his finance minister, his prime minister, the minister of health, and the ministers of education, transportation, and agriculture and say that this is what we have got to do to increase food production or to eradicate guinea worm, that you can make the program effective. That is not an easy thing to do. But I think that to the extent that this leadership group can put together a clearly understandable package, with my involvement and the involvement of many other people who can't be here, we can go to the heads of the nations and say that we have seen it work in Nigeria, Zambia, Sudan, or Ghana. This is what you will derive from it.

The attractiveness of this kind of commitment to a head of a nation is that it is one of the most beneficial political things that he could do. It is a very attractive thing for the people of Ghana to see on television several times a month Flight Lieutenant Rawlings out in the fields with the farmers, planting corn and harvesting a bountiful harvest of corn. Just recently, he went to eight different villages heavily affected with guinea worm and he learned enough about guinea worm so that he could speak [about it] with accuracy and authority. He tells the people of his country, to look at what is going on in this village, that all they have to do is strain their water and not drink the guinea worm eggs and they can do away with it. It not only helps his people but convinces them that he indeed is concerned about them. It is a very attractive political thing for these leaders to take active involvement in food production, health care, etc.

Olusegun Obasanjo:

Over the past four days, one of the things we have been looking for is thoughts from success stories in Africa. You have brought to us not one but quite a number of success stories that are happening on our own continent. This will add to the feeling of hope and the encouragement that we seem to be picking up here. You made one point that I think we should capitalize on: the fact that by virtue of the position you have held, by virtue of your standing both in your country and internationally, you can make those in high places listen to you. As these doors are made open, particularly in Africa, we will plead with you to impress on our leaders what you have just said: that we cannot have any progress especially in the area of agricultural production and food security unless we have their commitment. The commitment will have to be a total one. We believe, and that has been the consensus here, that the leadership in all walks of life is very conversant of the new awareness and what I call the new resolve in Africa. It is for us to take our own destiny in our hands and to change the situation. For that we need leaders who will be in front and whom we will follow.

Appendix: Summary of Main Conclusions

Conference by the Africa Leadership Forum, Ota, Ogun State, Nigeria, July 27-30, 1989

If Africa is to make the desired progress in agricultural production and food security and to prevent the impending calamity that the present situation portends, governments at the highest level will have to make the issue a priority one. Leaders need to show a clear sense of commitment in this direction and reflect this seriousness of approach in the allocation of resources to agriculture and food production programs and as a multisectoral dimension.

What Constitutes Food?

Food must not be seen just in terms of what is edible, but also in terms of its nutritional value, constituents, and what the people in a defined area consume for energy, growth, and sustenance, drawing essentially on the area's natural endowments or available resources and capacity. African countries must avoid the trap of reliance on imported foods so that food self-sufficiency can be defined in terms of the ability and capacity to produce or procure substantially within the region essential food items and constituents required by the people of Africa.

Food Security—National and Household

National food security should be defined within the context of national food self-reliance. It must imply adequate access by all people at national, local, and household levels to adequate and largely domestically produced food at all times. It involves regular and sustainable access without dependence on commercial and foreign-exchange–consuming imports or food aid detrimental to local production in Africa. In this context, food security should not hinder intra-African trade as a policy instrument; rather it should encourage and emphasize the full utilization of Africa's productive food resources. Increased consumption of imported food items, particularly grains and their derivatives, has been promoted by importing nations through overvalued exchange rates and through subsidies by exporting countries. The combined effect is that food import is made artificially cheap while local production is depressed as a result of price disadvantage.

Policies must be adopted at the national level to ensure realistic exchange rates and to discourage importation of essential food items through tariff and nontariff measures. Local production must be encouraged through price incentive and the encouragement of taste for locally produced food items. Donors, especially in nonemergency cases, should exercise restraint and caution in granting food aid that may discourage or depress local production.

What Went Wrong?

Several constraints of African agriculture were identified. These include the fragility of the tropical African soils, the blind adoption of technologies that are inappropriate to the African production environment, and the rapid degradation of the production system and environment. Agricultural research has been ineffective because in most cases researchers did not build on the practice and experience of the people and the farmers whose problems they were to address. Consequently, research did not fully incorporate the enormous amount of human capital embedded in traditional farming skills. Research recommendations that were inconsistent with traditional farming practices were normally rejected by traditional farmers who had knowledge and experience of the region and the crops and are normally skeptical of young educated researchers whose advice they regard as immature and fraught with danger. On the other hand, researchers have not been sufficiently motivated and as a result, national agricultural research systems have experienced a massive migration to the universities and the private sector, not to speak of those who went abroad to look for greener pastures and fulfillment in terms of remuneration and equipment and materials for research. Appropriate and affordable technologies have therefore not been adopted. Other constraints of national agricultural research systems include defective institutional structures that have discouraged active integration of research training and extension services, the poor funding and staffing of research institutes, and the failure to effectively disseminate even the limited research results that are available. The gross neglect of rural infrastructures and rural development has resulted in fragmented and agricultural production and agricultural markets, heavy postharvest losses, and a general strangulation of production. The limited research results have not been able to reach farmers because of inaccessibility and price incentives have not reached the farm gate because of numerous market intermediaries. Poor and woefully deficient rural road networks affect harvesting and transportation to market.

Inappropriate policies have strangulated African agriculture. These include failure to define the appropriate roles of government and the private sector in African agriculture, cheap food import policies, cheap urban food policies, heavy taxation of export crops, and unfavorable international trade policies, including unfavorable foreign exchange rate policies. African farmers have been victims of defective agricultural import and service delivery systems with the result that farmers do not receive inputs and services in the quantities they require and at the times that they need them most. Agricultural extension services have virtually collapsed and farmers are unable to generate adequate marketable and marketed surpluses. A particularly serious defect is the lack of institutionalized farm credit for the small-scale farmer because this has adversely affected the adoption of innovations and the expansion of the scale of farming operations.

Policy inconsistencies and instability have sent confusing signals to African farmers. Sources of policy instability include political instability with its attendant high turnover of the political and professional leadership of the agricultural establishments and the lack of consensus on the basic values and elements of agricultural policy. When agricultural policy proceeds in fits and starts, the private sector—the source of most agricultural investments—fails to receive clear and consistent signals that are essential for agricultural investments, capital formation, and sustained long-term planning.

Institutional proliferation has produced role confusion, overlaps, unhealthy rivalries, and ambiguities. The limited budgetary allocations to agriculture have been squandered on overheads of public parastatals and numerous public agencies. As a result, little or no services get to the farmers.

Particularly troublesome is the persistence of policy and program mistakes in African agriculture. The puzzle that must be addressed is why African agricultural policies repeatedly continue to produce so-called unintended consequences and unintended beneficiaries from year to year, from one government to another, and from one African country to another. The repeated failure to learn from past mistakes suggests that gainers from mistaken policies are politically more powerful than losers from such policies and that an understanding of the persistence of the African agricultural crisis and food insecurity must examine the power relations between the gainers and the losers from particular public policies in African agriculture. This emphasizes the need to develop and support widely organizations of farmers and rural people to exert needed pressure on national policymakers with a view to demanding better performance and policy changes so as to satisfy the interests and aspirations of the rural majority.

Another major constraint is the inability to control our population increase. Africa is multiplying its population faster than its food production. While food production increases annually by 2 percent, the annual population growth rate is 3.2 percent. African leaders must be sensitized to focus on the need to take urgent steps to remedy this situation. They must also take urgent steps to arrest the rural urban migration syndrome, which has had a disastrous effect on developmental efforts in Africa, but particularly in the area of food self-sufficiency.

Finally, there has been a woeful neglect of irrigated agriculture especially of the small- and medium-scale type through which Africa's dependence on rain-fed agriculture could be reduced, thereby promoting increased African food production in the process.

Women in Agriculture

The contribution of women in African agriculture has not received adequate policy recognition and accordingly the special constraints faced by women, especially in the denied or very limited access to farm credit, land, and production inputs, have not been addressed by policymakers. In some instances, women account for no less than 50 percent of the food produced on farms, but are only allocated 1 percent of the resources. Consequently, their full potential in accelerating agricultural production and food security has not been realized. The introduction of new agricultural technologies has not always been gender-neutral as some new innovations have unwittingly imposed economic losses on women in agriculture: the failure of researchers and inventors to incorporate knowledge of existing traditional food production, processing, and marketing between men and women. The marginalization of women must be corrected not only through sensitization of African policymakers but also through the active encouragement of women's organizations and groups. Policymakers must incorporate knowledge of intrafamily allocation of time and intrafamily distributoin of income and the impact of new technologies on the equitable distribution of intrafamily incomes. There is need for political and legal action to remove the traditional, cultural, and religions-reduced discrimination against women, which result in the constraints mentioned above.

The Small-Scale Farmer

The small-scale farmer plays a central role in the agricultural production process. Accordingly, he must be placed at the center of policy and its implementation. Some government policies, especially at the level of implementation, have discriminated against the small-scale farmer. Yet over 80 percent of marketed surpluses and supplies comes from small-scale farmers. In most cases, policies and research failed to recognize the large stock of scientific knowledge embedded in the small-scale farmer, acquired through process of intergenerational transfer of farming skills, based on trial and error methods. Agricultural research has made limited progress partly because of failure to observe and learn from the traditional small-scale farmer whose production system is well adapted to his environment. Policymakers must rapidly move away from the traditional vilification of the small-scale farmer as being conservative and resistent to change and seek a better understanding and appreciation of the farmer's extremely ingenious adaptation of his production system to his environment. Agricultural technologies must therefore focus not only on intercropping and the smallness of scale but must more generally incorporate complete understanding of why farmers do what they do and how they can do them better, cheaper, or with reduced effort and tedium. Small-scale farmers need to be trained in integrated farm production, which would combine crop and livestock production, agroforestry, fisheries and fish farming, water development, conservation, and management.

Agribusiness

An integrated view must include agribusiness, defined broadly to include not only processing of farm produce but also the upstream activities of farm input production. International agribusiness cannot solve the problem of African agriculture but it can make a useful contribution in the following areas:

(1) In the development or expansion of processing plants to handle farmer's crops, with or without an associated nucleus state. An important factor is the ability (and self-interest) of an agribusiness company to offer an assured market to farmers, as well as credit and extension services if these are needed;

(2) In large-scale crop production, as technical and financial partners for local landowners (or landowning companies). This may be particularly relevant in crops that are unattractive to small farmers, e.g., because of high costs, long maturity periods, or abnormally difficult physical conditions.

Policymakers must ensure that agribusiness establishments do not exploit the small-scale farmers/outgrowers especially when the agribusiness concern is the sole buyer of the produce as is the case with tobacco and sugarcane.

Subsidies

There is hardly any country that is self-sufficient in food without history of substantial subsidies on agriculture. African countries must devise carefully packaged subsidy programs that are consistent with the carrying capacity of their resources, noninflationary growth, and fiscal balance. Subsidies must be effective in relieving the famers and consumers and must not fall into the hands of intermediaries between farmers and consumers.

Political Will and Commitment

Many African governments have not shown adequate political will and commitment to successfully push through programs of self-reliance and food security. Even where these objectives have been nominally declared, they have not been translated into program and budgetary priorities. Policies have been inconsistent, unharmonized, and discontinuous.

Human Capital

National policies have consistently neglected the development of human resources in African agriculture. The development of African agriculture depends fundamentally on its human resources and programs must be developed to release the innate energies of local farmers and to reverse the rural-urban migration. Managers for farms and agricultural projects must be trained and utilized for the job for which they are trained.

Linkages Between International and National Research

African governments must move to exploit the potential contributions of international research centers that address their basic food staples. In particular, the national research systems must exploit the research results from the international research centers and exploit the complementarities in basic and applied research, which exist between the national research systems and the international research centers.

Energy

The role of energy, especially renewable sources of energy, is equally important in the agricultural production process. The objective should be to make energy more easily accessible to farmers. Such sustainable energy should be based on the potentially renewable energy sources in Africa.

The Forum raised and discussed the issue of having all hands on deck in the context of proffering solutions to all African problems, especially food security and all who militate against it. The suggestion for the creation of a Council of Former Heads of Government was regarded as premature because it was observed that there was no critical mass of such men or women in quantity for such a Council to materialize. And it was noted that incumbent leaders where possible do not shun the advisory services of their predecessors when no political loss is envisaged by the incumbent leader.

Recommendations

The Conference adopted the following recommendations:

(1) African Heads of State and Government must be sensitized to the deteriorating state of agricultural production and food security in Africa. Therefore, the Chairman of the Africa Leadership Forum is mandated, as a matter of

urgency, to call upon the current Chairman of the Organisation of African Unity (OAU), President Hosni Mubarak of Egypt, with a view to intimating the necessity of convening a special OAU summit meeting on agricultural production of food security in Africa. Such a session should also address the issue of how continental decisions and resolutions by the OAU on national food security can be translated into concrete political and program action at the national level. To ensure continuity on this issue, it is also recommended that the Forum Chairman continue such contacts and discussions with the Chairman of the OAU succeeding President Mubarak.

(2) The Forum shall establish a high-level group under the chairmanship of General Obasanjo that will undertake visits to a number of selected African countries in order to meet with government leaders and functionaries, research institutes, nongovernmental organizations (such as chambers of commerce, agriculture, and industry), banking institutions, women's organizations, farmers', and other organizations to sensitize them to the grim situation in African agriculture. This two-pronged approach should reflect the fact that African agriculture and food security are too important not to have all hands on deck and that the spectrum of policy dialogue must be broadened to include nongovernmental organizations.

In their meetings with African heads of state and government, the group should elicit practical commitments from these leaders with a view to facilitating a green revolution in Africa, to combating environmental degradation, to providing economic incentives to accelerate food production, and encouraging agro-industries and rural infrastructural development. It is further recommended that the Group should have consultations with the OAU, the Consultative Group for International Agricultural Research (CGIAR), the African Development Bank (ADB), the United Nations Economic Commission for Africa (ECA), the Food and Agricultural Organization (FAO), and the World Bank and its Special Program for African Agricultural Research (SPAAR).

(3) In order to encourage the free flow of information and communication among African governments on the one hand and between scientists and farmers in different countries on the other hand, a documentation facility should be established by the Forum, if not in Ota, then in one of the existing institutions in Africa to serve as a repository for technical and other information on useful comparative policy and program experiences and results of research. This information should be stored in such a way that it can be accessed directly by individual countries or serve as the parent stock for networking arrangements between different countries. The Special Program for African Agricultural Research (SPAAR) should be closely involved and consulted on mechanisms to access available research.

(4) A high authority on desertification should be established to coordinate the fight against the relentless advance of deserts in Africa. This authority should work closely with existing organizations now engaged in fighting or combating desertification in the various parts of Africa.

(5) The crucial role of women in agriculture must be fully recognized and African governments must be sensitized to the special constraints of women, especially with respect to denied or limited access to credit, land, and other production inputs. While it was considered desirable to pursue the establishment of an African trust fund for women as a special window of the ADB in addition to the resources that may be made available on the global level through funds like UNIFEM, it was felt that practical problems might make it advisable initially to concentrate on sensitizing national governments in Africa to facilitating access to more credit facilities for women.

(6) African governments should establish subregional food security systems as a nucleus of subregional and regional food security systems within the context of intra-African trade commodity flows. In order to avoid past policy mistakes, the reasons for past failures in establishing such groups within Africa should be frankly addressed.

(7) Financial institutions can play a crucial role in the promotion of agricultural production and food security in the region. While the African Development Bank, in its lending policies, already attaches high importance to agriculture, measures should be taken to realize a raise in the share of ADB's rural loan portfolio to 50 percent, with separate credit targets to be established to seek more ADB funds. To ensure the availability of the necessary funds, the international community should agree to a replenishment of ADB and IDA at a higher level.

(8) Given the critical role of rural infrastructures and rural development in attaining food security, the Forum should sensitize African heads of state and government as well as multilateral and bilateral agencies to the urgent necessity for a rural infrastructure and rural development initiative in support of agricultural and food production, marketing, and distribution. This may increase agricultural production, reduce postharvest losses in transit and storage, and encourage the participation of the private sector in the distribution of agricultural services.

(9) African governments should accord high priority to agricultural research extension and training as instruments for attaining food security. While national systems must rapidly build up their capacity for generating agricultural technology, accelerated efforts must be made to utilize available technologies. New institutional arrangements for generating and disseminating agricultural tech-

nologies should therefore be urgently explored. In particular, there is a need to vastly expand national capacities for adaptive agricultural research and to produce technologies adapted to Africa's highly variable production systems. A particular institutional innovation to be emulated on a broader scale is the establishment of universities of agriculture in African countries to ensure effective integration of research, extension, and training. Such universities of agriculture should spearhead new approaches to agricultural research, training, and extension as they have effectively done in countries such as India, Pakistan, Bangaldesh, Sri Lanka, the Netherlands, the United States, and Sweden. In this way, relevant and well-adapted agricultural research can be used as a potent instrument for national food security. Agricultural universities can also be charged with the reform of the education curriculum to emphasize agriculture at all levels of education.

(10) The Forum draws attention to the potential usefulness of establishing a technical aid corps for intra-African technical cooperation. But the immediate concern should be focused on the formulation of a coherent strategy for the reversal of the brain drain in Africa. One way may be to assign African specialists/technical assistants to other African countries.

(11) The Forum should also sensitize African heads of state and government to the need to mobilize the African youths for accelerated rural and infrastructural development. Farming needs to be encouraged and popularized, especially among the young.

(12) There is a need for harmonization of research findings and products from the work of extension service staff and the farmers.

Index